THE I-35W BRIDGE COLLAPSE

THE I-35W BRIDGE COLLAPSE

*A Survivor's Account of
America's Crumbling Infrastructure*

KIMBERLY J. BROWN

POTOMAC BOOKS | *An imprint of the University of Nebraska Press*

© 2018 by Kimberly J. Brown

Michael Dennis Browne, "Shall We Gather" from *The Voices*. Copyright © 2015 by Michael Dennis Browne. Reprinted with the permission of The Permissions Company, Inc., on behalf of Carnegie Mellon University Press, www.cmu.edu/universitypress.

All rights reserved. Potomac Books is an imprint of the University of Nebraska Press.
Manufactured in the United States of America.

Library of Congress Cataloging-in-Publication Data
Names: Brown, Kimberly J. (Journalist), author.
Title: The I-35W Bridge collapse: a survivor's account of America's crumbling infrastructure / Kimberly J. Brown.
Description: Lincoln: Potomac Books, an imprint of the University of Nebraska Press, [2018] | Includes bibliographical references.
Identifiers: LCCN 2017043083
ISBN 9781612349770 (cloth: alk. paper)
ISBN 9781640120693 (epub)
ISBN 9781640120709 (mobi)
ISBN 9781640120716 (pdf)
Subjects: LCSH: Minneapolis Bridge Collapse, Minneapolis, Minn., 2007. | Traffic accident victims—United States—Biography. | Brown, Kimberly J. (Journalist) | Bridge failures—Mississippi River—History.
Classification: LCC TG25.M33 B79 2018 | DDC 363.12/5092 [B] —dc23
LC record available at https://lccn.loc.gov/2017043083

Set in Lyon by E. Cuddy. Designed by N. Putens.

For Rachel, who makes all my days lucky

To live in hearts we leave behind,
Is not to die.

    THOMAS CAMPBELL (1777–1844),
    "Hallowed Ground"

Contents

List of Illustrations *ix*

Preface *xi*

The I-35W Bridge Collapse *1*

Acknowledgments *233*

Shall We Gather *235*

Remember the 13 *237*

Bibliography *239*

Illustrations

1. In shock, waiting for help *xiv*
2. Hood up *9*
3. Silver Saturn *10*
4. In the rescue boat *12*
5. Bent U10 West gusset plate *40*
6. Bridge 9340, October 18, 1993, inspection report *41*
7. Broken bolt *55*
8. Bridge wreckage seen through chain link *61*
9. Broken bolt and bearing block rotated *73*
10. Impact map *86*
11. Bearing block still rotated *95*
12. Abandoned bumper sticker *136*
13. REMEMBER THE 13 bumper sticker *142*

14. Pier held up with straps *155*

15. No repairs for rotting guardrail *156*

16. Cheap fix for Cayuga Bridge *157*

17. Another rotting pier under I-394 *159*

18. Bridge maps *162*

19. Repaired guardrail *216*

20. Repaired pier under I-394 *217*

21. Another repaired pier under I-394 *218*

22. No more straps *218*

Preface

When people learn I was on the bridge, they often tell me stories of their own close calls averting the bridge collapse. *I cross that bridge all the time—so lucky I didn't that day. My family was so relieved—I made it across. I was supposed to cross that bridge, but I was home sick. I was out of town. I had just crossed the bridge.* That these stories are so common makes it clear that Minneapolis's catastrophe is a shared collective wound.

The question *for me* when I began writing was What could've been done? Tears pool behind my eyes when I think about my informed view on America's breaking bridges. They pool because it's both remarkable and ordinary. Remarkable that I lived; remarkable the person I was before, so innocent; remarkable all the people whose lives were changed in thirteen seconds; remarkable the people I met, the abruptness of grief and shock, the effort to find positivity despite overwhelming horror, confusion, and fear. And ordinary: the warning signs that were ignored that could've prevented this man-made disaster; ordinary the way we live with disrepair. (Must we?)

So—so what that my bridge collapsed? (If I can briefly call it mine.) This question has been at the crux of my obsession, my drive to understand, to react and to act. Because what we don't do today will hurt us tomorrow. Because as Senator Amy Klobuchar said post-collapse, "A bridge shouldn't just fall down." It's taken nearly ten years of research, but now I know the answer to the question What could've been done? Presented in continuous sections like an unbroken bridge deck, this is the story of what happened to me, how it changed my life and everyone else's, and how it didn't have to happen.

THE I-35W BRIDGE COLLAPSE

1. In shock, waiting for help. Picture taken by author. © 2007 Kimberly J. Brown.

1

August 2007 in Minneapolis, the weather was tropical with dew points in the seventies and temps in the upper nineties. That summer I turned thirty-six. I was working out regularly—running, playing soccer.

In my house, enjoying the air-conditioning, I leaned back on the couch, squished some pillows to form a ledge, and flopped my leg atop it so my broken big toe would be elevated above my heart. After elevating for a while, I'd resume icing. I had filled a bowl with water and ice. Gingerly, I stuck my toe in the frozen sea. *Oh, that's cold.* I had learned from my big sister, Shelley, that you don't put ice *on* the toe, you put your toe *in* the ice. *Oh, that works much better.*

A year earlier a physical concluded that my blood chemistry and extra weight at the time had put me in a high-risk category. It was the perfect impetus to get my old body back, one kept fit from high school gymnastics and collegiate athletics. By summer 2007 I'd lost weight in the double digits, set goals, and had even begun running. I had run several 5K races like the "Get in Gear," Minneapolis's annual rite of spring.

Looking for a social outlet, one that would be healthy and active, I next decided that I wanted to try playing soccer. I had watched my nieces play, and they always appeared to have so much fun. Although I'd never played soccer before, I had once been quite athletic. I figured I'd give it a try, see if I could pick it up—though I had never played on a team and certainly had never played a whole game.

I remember going to a store called Planet Soccer to buy gear: those long soccer socks, shin guards, cleats, an actual soccer ball. How fun to put all this stuff on and practice dribbling. I kicked the ball for my dog, Lucy—an Australian cattle dog mix. She barked and gleefully chased the ball, batting it with both front paws, gnawing at it. When I was ready for more, I went online to a website called meetup.com and found out about impromptu matches on Sunday afternoons. Though self-conscious and unskilled, I tried to blend in to pickup games. One pickup game I remember was at a community field in south Minneapolis. The ball only came to me twice, my foot barely touched it (once, missing), but I was outside getting fresh air, watching, learning, running, laughing.

So, when a friend of a friend said her team, the Bombshells, needed players, I, having almost no experience, having barely kicked the soccer ball a few times during pickup games, joined. During the first practices, watching the other girls dribble the ball and participating in the soccer drills, I soon realized I was in over my head, but by then we were already short a player.

When I told my sister Shelley, I heard the surprise in her voice. "So you just decided to play? You know these girls have played probably since they were little kids." I said I knew and noticed they were all younger than me.

One teammate I met that summer, Kelly Kahle, lived only two blocks away, so we began carpooling. When Kelly asked, "Are you ready for two forty-five-minute halves of soccer?" I saw the shock on her face when I responded cheerfully, "Oh, is that how long they are?"

We talked rules, how to throw in the ball during side-outs, what happens during penalties. Commenting on the way I froze at the sight of an opposing player beelining toward me, a teammate complimented me on my ability to be a wall. My dribbling skills being nil, I probably should have quit. Three games before the end of the season, the opposition and I fought for the ball, and I got my foot stomped.

Before seeing a doctor at Tria Orthopedic Center, I felt dread, anticipating my dad's brand of medicine. I could just hear him. *I can take care of that loose tooth for you.* His brand of medicine involved a piece of string and a doorknob. Bam, no more loose tooth. Like jamming a shoulder back into its joint is how I imagined it. The doctor, donned in white coat, would yank my toe back into place.

The doctor put my X-rays up on the lighted board and confirmed that I had a fracture going up the lower part of my toe and another fracture into the joint.

I asked, "What about the way my toe was bent to the side?"

"It'll need to heal that way," the doctor said.

At first I didn't like how that sounded. Deformed! But then I felt relieved. All that was left to do was wait for time to heal it. I hopped out of the appointment wearing an orthotic boot. With an athletic background in gymnastics and tumbling, surprisingly, it was my first broken bone.

I couldn't drive. My car was a five-speed, and it hurt to push in the clutch. So I'd been working from home, connected by wireless. Look at the upside, I told myself. How great to skip the long commute.

But I was bored. I had become used to running every day or at least walking Lucy. Being sidelined . . . eating normal meals but no exercise . . . I worried I would get fat.

A week later, on August 1, the team would play its next game in Maplewood, a northern St. Paul suburb. Though I couldn't play, I debated going, just to watch.

After much internal vacillating, I decided it could be fun to go. Get out of the house. Play team photographer.

My wife, Rachel, dropped me off at Kelly's summer job, where I'd hitch a ride. Kelly poked her head out the heavy wooden door. She was at a church where she worked with youth.

"Almost finished."

"Okay, no problem."

Waving to Rach as she pulled away in her beige Mazda Protegé, I settled on the short steps and sat on the stoop. For no particular reason (Rachel told me later), she felt apprehensive about leaving, but I told her I was fine—no need to wait. I put my chin on my palms and watched the street. A tree-lined Minneapolis neighborhood, an old street with rows of duplexes, a peaceful few minutes.

2

The poet Langston Hughes asked, "What happens to a dream deferred?" Does it "fester like a sore" or "stink like rotten meat?" What happens to maintenance deferred? "Maybe it just sags / like a heavy load." Or does it implode?

It's one of those hot humid days where just sitting still, my nose begins to sweat. The heat tops 90 degrees, the heat index higher. The air is like a dense wall.

August 1, 2007, evening rush hour. We are crossing the Mississippi River in Minneapolis on the 35W Bridge. Ride along with me.

Kelly drives her silver Saturn. I ride as a passenger, unaware that we've started across a bridge. We head north on 35W—a major freeway, part of the Interstate Highway System that slices down the center of the United States from Duluth, Minnesota, to Laredo, Texas. Before me, the car dash smooth and sturdy, a clear view out the passenger window frames the city. I don't really notice the cloudless sky. I watch red brake lights brighten. We approach, gradually stop. Behind us I see cars in an endless stream in the side door mirror. Cars, pickup trucks, and semis grind to a halt. Stop and go. In a construction zone, two lanes are closed in both directions. My hands pass the time fiddling with my cell phone, its safety strap loose around my wrist. Scan the lighted pad, the picture of our dog, Lucy, as I hunt and peck sentences. I wear a teal Nike sleeveless top with a collar and denim blue jean shorts—short, bunched at my hips, curlicue embellishments on the side. My bare thighs are tanned, muscled from a summer of running. I had stashed a camera and a thin wallet in my back pockets.

The line of cars advances. The space between cars grows. The world outside, the road, with its white lines and yellow hashes, fills with work trucks, stone piles, and construction workers dressed in jeans and reflecting vests. They strut across the expanse, lips moving. In the closed car I notice the faint scent of soccer gear, as Kelly switches the air from fan to AC. I turn the vent slats, point the cool air at my face, which beads with a fine layer of dew. The road pulses, and the car bumps as we roll over seams. We creep at five or ten miles per hour. My upper body wavers with traffic. Then my head stills into the headrest. My big toe throbs, pulses like the road. The seatbelt occasionally tightens across my chest like a pageant sash. I lean forward to type more into my phone. Who I was texting, I don't remember. Deaf to the drone of traffic. Blind to the churn of the Mississippi. Clueless that 114 feet down, on the river's surface, foam swirls in a froth like cappuccino. I know I'm pointed north. That the driving is in a friend's capable hands.

Then I feel the road shake. Jolt from texting. Again, stronger. Feel the violent jar, the pulse of something that can't hold. Whirr my gaze to the windshield. Rumbles turn to waves, and the flat bridge deck breaks. It rollicks like the Atlantic, Pacific, Indian Oceans. Gyrates like the platform of a Tilt-A-Whirl. I tell myself I must be wrong. Witness the world wane.

Kelly yells, "The bridge is going to fall!"

A split second—I don't speak. I think, *No No No No!* This can't be it!

The car tips askance. We shimmy like a carnival ride, except we aren't safe at all. We roil like popcorn kernels in an air popper, vibrating helter-skelter. My hands fly up, in my small private space, to hold on. The road coughs—a quiver, a rumble, nauseous, the dynamite before it blows. My body seizes. Breath stops. Muscles light afire. A stream like poison races to all points. Terror in veins, the burn of a chemical reaction spurts as the threat soars to the brain and registers.

No tears. Life doesn't flash before my eyes. Only a rise in my throat, my screams. Lungs blast like the AC as I fall with the bridge. The energy rattles through my chest. The momentum slows at fractional moments crashing in rhythm. Boom! Boom! Boom! Down. Down. Down.

The car sinks. My body attempts futilely to lift, suddenly heavy. Gravity sucks me down. The faint presence of the seatbelt expands without my awareness as I bend forward, crank my arms up, and jam my laced fingers behind my head. Falling momentum yanks yanks yanks. My stomach is a rock hardened by fear. We slam into obstacles—catawampus, crooked, nonsensical directions. Legs loosely flop like a marionette.

I implode with the bridge.

Lift, mash, and convulse with the road.

Sense my tininess

as I fall

in

the

little

tin

box—

Trying to shield myself, I close my eyes, grab my head and neck. I scream in terror and panic, "Fuck! Fuck! Fuck!" with every crash.

My chest repeatedly sears into my knees as the fall tosses me akimbo.

All those cars in a row fall too. So do the stone piles, the temporary concrete dividers, the trucks with their cement drums turning like globes, the road signs, the microcosm of my lap, my phone, Lucy's image tucked under the closed lid, all the phone numbers and names I'd stored. The falling world doesn't just clatter—it booms, repeated booms that deafen and zing ears. The concrete fractures. Steel scrapes and roars.

From inside the vortex I am Dorothy with her dog, Toto, launched off the ground. I will be crushed. I will be ended. Crashes so encompassing, I cannot tell where they are. A sound like garbage dumpsters, when trucks lift and drop them on the street. Loose in the air, the crashes mean something massive, out of control, might land on me. The windshield washes from clear to mud with debris and colors of beige, coffee, tan, taupe, oak, yucca, wicker, meat, turkey. Then the expanse turns solid like a mudslide. Shattering glass singes memory. The light is gone.

There is no light. There is no light.

I slam my eyes closed. Blind myself. Split seconds! Between larger crashes, smaller cracks ding against the windows and the roof. Cement, cars, people, and roadway. A monster's yell. Something gigantic will smash me to smithereens.

Grit my teeth, tight.

Tighter.

Tighter, I said.

Bounce with the fall. Hold my breath. When will this be over? Let it be quick! No time to say, "You're okay," my prayer in rough skies when flying. I anticipate the end, the crunch of road landing on me. I will break in two, flattened, bones crushed into powder. How many times have I squished spiders? I'll be unrecognizable. Ended in a moment. Blink, done. Or it will be longer, partial. Broken but aware, I will be a shaken spatter of paint not drying immediately. I'll scream in a vacuum, in the depths, under this inhuman road. Kelly's screams sound distant. I scream in a pitch and tone I don't recognize: loud, shrill, tinny, losing oomph, losing breath, losing.

Stomped to the earth, ground out like a cigarette butt, we land. Dead—no wait, alive! I can't believe it, but there's no time for that. River water breaches car joints, pours into my lap, and begins to fill the car. It stinks like bad breath and shocks my skin. Wrong, wrong, this is all wrong! Open eyes. Assess. Now the water feels lukewarm. This isn't over! I recline like an airline passenger at takeoff. Arms and legs weigh a hundred pounds each. I must be sinking. Start the scramble to survive. Move, dammit! Hands accomplish nothing. Is it over? Something could still drop from the sky and crush us. Yell to Kelly, "Drowning! We might drown!" What do we do? "Get the car windows down! Get them down!" Press the button on the door and look to the window. The lock mocks me, as windows stay put and the lock sings click-click, click-click.

Kelly says, "This stupid Saturn! The controls are in the center!"

Reach for them. Hope: please unroll. Push on the controls. Wrong! Pull on the controls. Right!

Like a stage curtain, water recedes from the windshield. Darkness in the cabin peels back like a camera shutter, a sick joke. The water in the car shrinks. Later we find out why. The flooding came from the splash of the bridge deck slamming onto the river. Unclick the seatbelt and open the car door to nothing. I gaze downward from the open passenger door to the water below. I want to escape out my side, but with the car tipped, hood up, it's a distance to the river's surface, which sloshes below. A thick steel beam, a smooth curve of green, disappears into the depths. Meanwhile, Kelly-turned-Houdini has, unbeknownst to me, pulled herself out of the driver's side window, and now she stands in front of the car, on the fallen bridge on the river.

I yell, "I can slide down!"

Kelly yells back, "No! You gotta go out my side!"

"I gotta get out!" I yell, estimating the distance from open door to water. A far drop, maybe as far as I am tall. A decent jump, but I can do it. Wiggle forward. Grimace. Wiggle forward, dangle feet. Slide before the car sinks. What am I waiting for?

Decide.

I must go out the driver's side window. Hesitate. The window beckons, miles away. Move, and the car might slither into the river. Drown me in its filth. Go anyway. Scramble on the balls of feet. Scale the seat creases. Squat.

Kelly says, "Do you want to try to get the car into park?"

"There's no time," I say, "I gotta get out of here!"

"Yes, right!" she says, as I stick my foot out the window. As I grip the doorframe, my cell phone dangles from the safety strap around my wrist. I yank it off, but the strap catches under a heart—or rather, the shape of a heart—on a ring that Mom had given me. I balance on the window threshold. Nowhere to step on this side either. I stretch my leg toward the bridge edge, just out of reach. Green sticks, what I later learn are rebar, poke toward me from the broken bridge edge. Toe them. Graze them. Wiggle my fingers into the line, the slightest gap, between the hood and side body panel. My fingertips don't fit, but there's no other option. Pull.

Say, "There's nothing to grab! What did you grab?" *Do not stop worrying about the car slipping back.* If it does, I will fall. Make squeaking sounds but keep moving. Grope the windshield. Grasp the wiper. Feel it yank back.

Pull, tiptoe, repeat.

Kelly says, "You're going to do it." I grab, look, readjust. Tiptoe higher on the rebar. Wiggle fingers deeper into the hood seam. Take Kelly's hand.

Alley-oop!

. . . onto the broken bridge, whole sections of bridge deck that landed in the river and are now visible above water. Clutch Kelly, who I barely know, around the neck. Bury my face in her shoulder and howl a single punctuated scream.

Separate.

Stand speechless.

Feel the awful adrenaline scourge my body.

3

Kelly and I stood on concrete, but we were stranded on an "island," the broken bridge in the river. "No . . . ," I whispered.

Concrete chunks littered uneven surfaces. The bridge deck rent asunder, like land shifting after quake, fissure of earth fracture, pressure overcame strength until the material ripped, ignoring seams, ignoring rules. On land, past shore, concrete tore, and steel folded and bent in a mishmash of tangled green. The road lay quiet, eerie silence. Kelly's front car tires spun, and exhaust mushroomed from the rear. The organic decay of rotten fish combined with chemicals—the pungent unmistakable odor of fuel. Waves sloshed as swaths of black smoke drifted downward from above, where the bridge had shattered across the frontage road. As I walked the slanted broken bridge deck, tangled sections of roadway alternately curved and disappeared under water.

While we waited, Kelly and I took turns taking pictures with my digital camera, which survived the fall shoved in my back pocket. Waiting for help, with nothing to do but stare at circumstance, I lifted my camera, arm extended as I'd done countless times with Rachel, and snapped a self-portrait. Muscle memory lined up the shutter. I wouldn't know it then, but later I'd want this physical artifact. Proof that I was there.

Then the quiet disintegrated into cacophony. Fumes burned my nose and eyes. I called Rachel. I willed her, pick up, pick up. Her voicemail said, "Sorry I can't take your call. Press 1 to leave a voice message." I pressed 1. "Rachel, the fucking bridge fell. I'm in the river. Call me."

2. Hood up. © 2007 Kimberly J. Brown and Kelly Kahle.

4

I held my cell phone for a curly-headed red-haired man who clutched his T-shirt-wrapped hand at his chest. People began to emerge from their cars. Helicopters arrived and hovered overhead in increasing numbers. Cell phone signals were now jammed, and Rachel couldn't reach me. I called my sister Shelley and got through, maybe because she was on a land line. The whomp of helicopter blades and emergency sirens droned.

"What?" I yelled into the phone, straining to hear.

Watching TV coverage at home, Shelley said, "I can see you!" I looked skyward. She rattled off details, but a helicopter swooped overhead. I pressed the phone to one ear and covered the other. She described explosions and fires and warned that cars may have leaked gas.

"Move away from the cars—they're exploding above you. See the smoke?"

From the concrete island low in the river, I saw the smoke but not what was burning. I wavered between two cars, aiming for the exact middle. I said, "There's nowhere else to go! I'm ready to jump in the river!"

3. Silver Saturn. © 2007 Kimberly J. Brown and Kelly Kahle.

"No, don't jump in the river!" Shelley said.

Later I heard that a pastry truck had caught fire. Someone said, "There's a school bus up there." Rescue boats eventually approached. According to the time and date stamp on my pictures, the flotilla arrived around thirty minutes after the collapse. Rescuers boarded the "island" created by the bridge ruins in the river. A man in a yellow hazmat suit exited one boat and gestured at me to get in.

"Not without her," I said, shaking my head and pointing to Kelly. She didn't appear to be hurt, but I had seen enough movies. I'd be rescued, and then something would happen to Kelly and I'd have to live with the guilt. Besides, others were hurt worse, I told the rescue guy. He huffed but moved on to others. Another boat arrived with room for both of us. A rescuer handed me an orange life vest. I put it around my neck, looped the belt around my waist, cinched it tight, and snapped the buckle until it resounded with a satisfying click. A driver in the boat extended his hand. Once full, the driver steered the boat away from the wreckage toward an opposite shore. The high-pitched sounds of emergency tinged the air. Whoosh of helicopter blades. Incessant sirens from firetrucks, ambulances, and police. The rev of the motorboat engine. Kelly sat opposite me and took pictures with my camera. One showed my beige bra strap falling down my shoulder, past my teal tank top, my mouth ajar, gazing again at my phone.

The rescuers dropped us off at shore, near the base of the Tenth Street Bridge. At the time we didn't know the names of the bridges. People guessed, but no one knew for sure. We stepped out of the boat, and the driver said, "Ambulances are over there," pointing toward a dusty area by the collapsed north end.

We started up the rocky shoreline—me in my orthotic boot. Kelly held my hand to help me climb the boulders. Behind us the arches of the Tenth Street Bridge framed the mutilated 35W Bridge in the water. Once on the northern riverbank, we continued our climb, escaping the river. We ascended a steep incline, a long hilly blacktop-tarred road, toward the backside of university buildings, panting as we climbed. People on bicycles charged downhill to see what had happened.

"Don't get taken out by a bike!" I yelled as bikers flew down the hill and zipped past.

From the hilltop the view expanded impressively. From the height of the university's east bank overlooking the rugged topography and the multiple bridges within that area: an abandoned railway bridge, the Washington Avenue Bridge in the distance, and the Tenth Street Bridge and the 35W Bridge running parallel. On a normal day hikers in this area would be enveloped by the sound of rushing water from the lock and dam, against the constant haw of crows. If you listened closely, one bird cawed, paused, then another bird, farther away, answered.

After scaling the hill, we turned left by a line of stopped graffiti-covered trains. Pebbles, gravel, and stones crunched under our footsteps. Lost in the underbelly of the university area, which way out? Railroad tracks crisscrossed creating a pathway of sorts, but it was wrong. The Dinkytown area and Sanford Hall, my former dorm when I went to school here, had to be nearby. My sense of direction, gone. A beaver or some kind of mole lumbered over the dusty landscape. To return to the south side of the river, we

4. In the rescue boat. © 2007 Kelly Kahle.

crossed a pedestrian bridge called Dinkytown Trail, formerly Northern Pacific Bridge 9—a steel deck truss like 35W had been. People lined up along the railing—their energy palpable. I looked briefly but turned away, nauseated by the gawking. The gathering people were becoming a crowd. Numb, I walked amid the tumult and the burgeoning chatter and wondered, Could they tell that I had been down there? Everywhere I looked, faces. None of them hers. Where was my Rachel?

5

When we reached the other side of Dinkytown Trail, on the university's West Bank, cops sauntered downhill waving their arms, shouting, "Get out of the way!" Some people were rude and talked back.

Get me out of here.

A surreal sense grew in a state I can only describe as cognitive dissonance. These events had *just* happened. Fresh images scrolled in my mind, trauma mixed with reality. Being rescued from the broken concrete in the river and taken to shore by the boats; limping up the rocky embankment; deciding I didn't need to stop at the ambulance as stretchers rushed past with seriously injured people on them; dodging cyclists rushing downhill to see the damage, me wanting to get far from it; walking on dirt roads next to graffiti-covered railcars; being lost; trying to figure out a way to a major road; water dripping from my clothes; broken glass clinging to my body.

Emerging from the gathering crowd, onlookers rushed toward the nightmare I'd just escaped. The mood turned to pandemonium, an exodus toward ruin with energy akin to rabid mice. I turned toward an apartment building called Riverview Towers, then toward Holiday Inn in an area known as Seven Corners at Cedar and Washington Avenues. A sea of faces. I turned left, strangers. I turned right, strangers. Their faces buzzed like atoms in a cell. I searched.

Question: Facial recognition?

Answer: Negative.

Then the moment. There! Rachel craned her neck, searching, from a spot on the patio of a bar called Town Hall Brewery, amid tables, chairs, and foliage. Her face emerged from the discombobulated mess. We rushed together with a smack, our mouths pressed together hard—urgent, chaotic, dire. Sloppy tears wetted our breathless kiss as we embraced. Faces melted away as we held, shook, and wept. Overwhelmed, I shut my eyes. Her eyes, mouth, skin. The jut of her shoulder blades when I clutched her. The gravity of her grip. No withholding, none of the usual carefulness about our public display of affection. Not then. I'd just cheated death.

After more than an hour of walking from the river, I finally found her. My Rachel!

We walked to the car, dazed, arms encircling each other and thinking, this couldn't have happened. People ran, yelled, gawked. Cops blew whistles. A disordered scene of humans bleating all around us. We advanced through the throngs, crookedly and awkwardly, amid the screaming sounds of emergency, consumed at the shock to still be alive. Clinging.

6

Rachel drove Kelly and me away from downtown. First, we drove to an athletic field to try to find Matt, Kelly's husband, who was playing Ultimate Frisbee. We found an empty field.

We headed home, to our house. When we arrived, the view overwhelmed me. Tromp in the back door, step up a stair onto the green linoleum kitchen floor. White cupboards, blond butcher-block counter, overhead storage cabinets. Above the stove, open space overlooks the dining room and our farmhouse pine table. Past there, Klik-Klak couches and chairs.

In a wide-eyed, staring, disoriented state of shock, I noticed things I wouldn't normally. The widow's peak on Lucy's forehead where her black fur changes to bronze. Color. Home's walls never seemed as richly luxurious as the moment I first returned. Crisp lines of walls, window frames, coved ceiling and doorways. A nearly undetectable hum of house sounds: a murmur, air escaping a floor vent. The levelness of hardwood and perpendicular angles of borders that make sense. Although I physically stood inside the threshold of my own house, I had a feeling I can only describe as a duality of existence. I watched myself, above myself, only partially in my body.

We turned on the TV. There. Seeing the disaster on TV is the only way to sense the big picture, which in person was incomprehensible and gigantic. Unbelievable, the news. Cameras pan entire shots of massive lengths of wreckage evident only on that screen. The middle part in the river. The north end, crumpled and fallen on dry land, crushing a train car carrying some kind of grain. The southern part fallen in graduated levels—lower and lower—from the upper part by Washington Avenue down across the frontage road. Later, it will be discovered, a nearby surveillance camera recorded the debacle. News stations will play it on a slow-motion loop. The footage shows the deck flex, plummet, and then smash into the river. Water shoots skyward like a surfer's perfect wave. The bridge's north and south ends crumple over their piers. The center span tops the river surface—the deck broken and jagged, buoyed underwater by thousands of pounds of crumpled steel. At the time I couldn't tell how the car stayed above the water. In photos and on TV coverage we will see that Kelly's car seems to hang from its front wheels from a broken bridge edge, the lower end of the car submerged in the river. Meanwhile, people strewn topsy-turvy in cars have crash-landed.

Totally unsure what to do, I paced our house. Take off these clothes? I trailed bits of glass from the door, through the kitchen, by the couch, into the bathroom. Legs covered in bloody cuts, some clotted with thick bulbous gobs, I sat on the ottoman for a second. Glass shards fell on the floor. What do I do now? Keep clothes on? Try to take a shower? Go to the hospital?

Later I opened my phone and saw an *I* for incomplete delivery. I had sent Rachel a text at 7:03 p.m., not quite an hour after the bridge collapsed. She never got it. I had written, "Were [sic] ok."

My sister-in-law Donna asked if I'd interview with Diane Sawyer on *Good Morning America*. Her friend Sandy Lynch worked for ABC and was asking. I said yes, feeling detached. Being asked if I wanted to interview for a major national morning news show was more than unusual.

I continued trailing glass bits around our house and sat again on the ottoman. Seeing the collapse on TV made it easier to detach from reality briefly: the enormity curtailed by the fixed dimensions of a video screen.

Family encouraged me to get checked out at the hospital, so from our house we piled into the car a second time, and Rachel drove Kelly and me to the emergency room. Once there, Kelly's husband, Matt, arrived.

In the emergency room waiting area, the collapse dominated the news. TVs in the waiting room, mounted on walls in corners near the ceiling, all covered the event. People sat in the waiting room watching, and I wanted to tell all of them, "We were there!" On TV. I couldn't believe it. This kind of thing doesn't happen to me. We watched, and I talked with Dad on the phone, pacing, pointing at the images on TV, pointing at the car we were in, as if he were there with me at the ER and could see, as I tried to describe what happened. People later asked, "You were there?" I nodded. Surreal. How could this be?

I filled out forms and answered questions. Then I heard my name called. Rachel followed me into a room, part of the area separated by hospital curtains hanging from ceiling tracks. A nurse told me to lie on a gurney. Nurses led Kelly to a separate area.

"When was your last tetanus shot?" one of them asked. I shrugged. "Okay, we'll do that." A nurse swabbed my bicep with iodine and gave the shot. "Well, you'll never forget when you got this shot," the nurse joked. The lights seemed overly bright. He pulled on gloves. Thwack, as he adjusted them. He scanned my legs and inspected every inch, running his fingertips over skin, feeling for sharp spots. When he found glass, he plucked or rubbed it off or used tweezers to dig it out. In the other hand, he used a gob of gauze to catch blood and dab. I felt the localized stinging of each shard. Broken glass on the underside of my legs had stuck deeper.

"Sorry, won't be too much longer," he said as he dug. More dabbing and a stick to make the blood clot. Pain is good, I told myself. I welcomed it compared with what I had thought earlier was going to happen.

The nurse bent my knees to my chest, rotated my bent leg at the hip, one at a time.

"Does this hurt?" he asked.

I shook my head no. Sandy Lynch arrived, and I heard her talking with people outside the curtained area. Someone intercepted her. Most of their conversation was inaudible, except for the words *privacy* and *consent*. A male doctor in a long white coat poked his head past the curtain and explained that I didn't have to see the media or talk to them—it was up to me. It's okay, I said. Sandy entered, and our conversation reminded me of all the crime shows I'd watched on television. "Sorry to ask you this, but can you tell me what you remember?" I gave highlights summed up with "It was terrifying."

That night I walked from the ER on my own power, on two strong legs. I had cuts and bruises. I didn't break any bones. No casts, no crutches, no wheelchair.

We got home after 2:00 a.m. I staggered to the computer and emailed my photos to ABC and sent emails to family, close friends, and my work colleagues. (I'm a technical writer.) Then I lay down to rest my eyes.

At 4:30 a.m. Sandy would pick up Kelly and me for the interview. I didn't really sleep. After about an hour I got up. Might as well get ready.

In the shower my skin stung. Eventually, the stinging dulled to vague throbbing, and I figured my body was swelling to protect itself. Shampoo suds trickled down my neck, and as I soaped my earlobes, I found glass in my ear.

Oh my god oh my god. How do I get this out? Don't panic. Can I swab it out with my finger?

I touched it and felt the shard stick. Crap. I took the showerhead nozzle down from the wall. I leaned forward, turned to the side, and faced my ear toward the floor. Then I turned the water on low and, gingerly at first, sprayed the stream upward toward the side of my head. *Please don't let it get lodged in there, please don't let me go deaf.* Eventually, the glass washed away. Relieved, I got out and finished getting ready.

I picked out clothes, put on makeup, and when I was ready, I kissed Rachel as she tried to doze, and I said, "I love you I love you I love you."

In the living room I peered out the front drapes. Outside the streetlamp shone a bright circle onto the sidewalk. No car yet. Full of nervous energy, I decided to wait in the three-season porch. I gingerly closed the front door behind me and waited in the dark.

For the interview we stood on University Avenue at the entrance ramp to southbound 35W. Construction cones and fencing blocked pedestrians and automobiles from taking the deadly route down to the ruined bridge. Surreal, that morning at 4:00 a.m. on August 2, technically. The wee crisp hours after surviving a bridge collapse. News trucks lined the sides of University Avenue. Satellite dishes lifted their cupped faces to the obtuse blue of oncoming morning. Streetlights cast a lemony hue against dawn as it crept over the area.

While waiting, I listened to Rachel's voicemail from the night of the collapse.

"Hi-iii!!! Oh god, I've been trying to call you for so long, and it would never go through.... I've been trying to find you too. People keep telling me different places to go. So I'm headed toward the Red Cross Building at... 1201 Second Street or River Road, whatever, so I'm going to be there in one minute, and I hope you're going to be there soon too." Her voice catches. "I love you.... Bye."

I cannot delete this message.

Reporters and personnel from various TV outlets milled about, swelling to a small crowd, near the fenced-off area by the 35W on-ramp. *Good Morning America* interviewed four survivors: me, Kelly, Gary Babineau, and Melissa Hughes. Sandy Lynch waited at the sidelines as the GMA camera crew lined us up on the sidewalk. They rigged each of us with an ear bud and a microphone. They told us to attach the battery pack to our waistbands and snake the cord up underneath our shirts, then they clipped them near our collars. We got a thirty-second primer on how to interview. Basically, keep your answers short.

When the interview concluded, we stayed and talked with Sandy and the ABC crew. Then there was a brief time for people watching. NBC *Today* cohost Matt Lauer talked with Minneapolis mayor R. T. Rybak before cameras near a fluorescent orange construction sign. I scanned the crowd. There was *Today* cohost Katie Couric. I wanted to introduce myself and tell Katie that I watch her every morning and I had survived! But I felt shy—plus, I had promised ABC an exclusive interview. In retrospect that wouldn't have meant I couldn't have said hello, but I wasn't sure what might happen.

Later that day a producer from ABC *World News* called and invited me to interview with Charles Gibson in the afternoon. I heard myself agree to it with the same sleepwalking feeling as when I said yes to *Good Morning America*, as though I were separate from my body.

On our way to meet him, Rachel drove. I hadn't been paying attention. At some point I looked out the window and noticed the stop-and-go traffic on the Third Avenue Bridge, another crossing upstream from the 35W Bridge ruins. Ahead, at our left, at our right, cars stopped. Trapped, sandwiched between bumpers, panic flattened me. I whipped my gaze to the floor mats and held my palms near my eyes like a racehorse with blinders. Oh god. Hyperventilating. I was on another bridge. That was the first hint that I had post-traumatic stress.

Under a tree whose arms sheltered us from the pulsing sun, Charlie asked me and another man several questions. One sticks in memory. "Are you angry?" Surprised, I said, why would I be? I was so grateful to be alive. I wasn't crippled or disfigured. What else was there to feel?

7

In the United States, when a bridge collapse occurs, the failures provide critical information for engineers to learn from past mistakes. For example, the collapse of a New York State Thruway bridge in 1987 that killed ten people, alerted officials to the problem of scouring on underwater bridge supports. (Scouring occurs when swirling water erodes sediment around underwater supports.) Or the 1980 collapse of the Sunshine Skyway Bridge in Tampa Bay (rammed by a ship), which focused attention on the threat to bridges from water traffic. "Sometimes it takes a tragedy to get decision-makers to pay attention," Andrzej Nowak, a civil engineering teacher at the University of Nebraska said. "I think the biggest thing to come of this will be that, as a nation, we'll spend more money replacing and maintaining these older bridges." The collapse of the Silver Bridge in 1967 in Ohio, in which forty-six people were killed, led to the new National Bridge Inspection Standards, specifying that bridges would be inspected at least every two years.

The National Bridge Inventory rates the stability of bridges on a 9-point scale. The best score registers a 9 for "excellent condition." According to the NBI Rating Scale, 4s and 3s are in "poor" and "serious" condition—ugly corrosion, cracks or chips that threaten piers, superstructure, bridge deck, or substructure.

8

Bridges meld into the landscape, and eventually we cease to see them. They become invisible. So accustomed to seeing them, having lived with them, we don't recognize all the bridges we've crossed, accelerating over highway or crawling past one side of freeway to enter another. The average citizen lacks the language to describe them or the structural concepts to know how one differs from another. We want to see simply, see lines and know they lead the way. But bridges are complex. Their logical straight lines hide invisible forces that are as relevant as their physical presence.

A "bridge" is anything conveying traffic (of any kind) over another natural (lake, river, ravine) or man-made entity.

The roadway sits on the deck, which is usually made of reinforced concrete and occasionally of wood. Sometimes movable decks are composed of steel grids. The substructure consists of all parts that support the superstructure—the main parts being abutments, piers, footings, and pilings. The superstructure consists of the components that actually span the obstacle the bridge intends to cross. It includes the bridge deck, structural members, parapets (low structural walls), handrails, sidewalk, lighting and drainage features.

The first person who built a bridge across the Tiber River in ancient Rome was called "Pontifex," or "bridge builder."

Julius Caesar and the pope were called "Pontifex Maximus" for their role in connecting people. Bridges are engineering feats that can define an area or locale, draw tourists, or lure sightseers. In 1867 John Roebling dreamed up the first modern suspension bridge, the Brooklyn Bridge. In 1961 the Eisenhower administration built the Interstate Freeway system, and urban sprawl was born.

A bridge can be many things: a structure that helps us span a depression or obstacle; the part of the nose where glasses rest; the piece that lifts strings on a musical instrument; the forward area of a ship where a captain steers; the tool that stabilizes a pool cue for a tricky shot; part of a denture. A bridge can even be microscopic, a single atom: one molecule kissing another.

Close your eyes. What do you see? Darkness. Look ahead. There. Straight ahead at the underside of your eyelids. In your vision notice the absence of shapes. Can you describe this nothingness? Tell your brain to translate. Study the subtle static haze of micron-sized dots. Now: go into the invisible, into the invisible of a bridge. That's right. Go into steel, into girders, into twisty rebar and heavy concrete, up through piers and into the deck—become a force, not a being—and slither into the superstructure. Explore the underside. Span abutments, piers, pilings. Charge forward along the length of reinforcing beams. Feel invisible forces: compression, load, shear, tension, vibration.

Minnesota. State of water. State of extremes. Of snowstorms and humid summers replete with the state bird: the mosquito. We live in a place that fluctuates, a state that yearns for something it's not. In summer we say, "We're boiling!" as we roast under a cloak of tropical air. In winter we hide indoors, waiting for clemency from the frigid Canadian zonal flow that drops our backyards to minus 10 and below. We use various methods to quantify our discomfort. Windchill. Heat advisory. Dew point. "Feels like" temps. From minus 45 to plus 100, we lean upon the inexact science of description, comparing our feelings to air temperature. On a human level we unwittingly sense what steel, concrete, and rebar experience day to day. In the cycling extremes of Minnesota weather, bridges have their own ways of responding.

The 35W Bridge rises a short distance downstream from Mill City Ruins Park, a recreational area along the Mississippi River in Minneapolis. The ruins are named after a flour mill that once stood here and was destroyed by an explosion and fire in 1874 that killed eighteen. If you were walking here, you'd see grass freckled with clover. Maybe a kid gliding past on a skateboard, people coasting by on bikes or walking and jogging on pedestrian paths.

A mist of water vapor rises and blows in the air, obscuring the view across where buildings dot the other side of the riverbank. There is a building with long banners down its facade. ST. ANTHONY FALLS LABORATORY says one banner. Two others say EARTH and WATER. There's another banner, but the font and background are so light, I can't read it from this far away. Looking longer, I notice the University of Minnesota "M" logo. Cutting across the river, the Stone Arch Bridge. Other bridges are here, like the Central Avenue Bridge.

How is it you come to a place, you see it again and again over years and days, yet you sometimes don't connect the dots? You don't know the meaning that place will have in your life.

9

The day after the collapse, I urgently saw my family: brothers, sister, dad, sister-in-law. A friend, Joanne, made me a tray of cheesy glad-you-survived lasagna. Friends sent flowers and notes and emails expressing their shock and to congratulate me on surviving.

Media initially reported the fall as 65 feet, but that was only to the underside of the steel trusses. From the bridge deck to the river measured 114 feet. A factor that may have helped people survive: we were in a drought that summer. The river depth was a foot lower than usual. When the wreckage landed, it was speculated that the truss structure below the deck may have "cushioned" the fall versus if there had been no framework underneath.

In my replies to emails I said, "I don't mind losing at bingo, the lottery, or raffle drawings. I used up all my luck."

Shelley emailed our pictures to her friends and wrote, "By the grace of God she and her friend are unharmed." Her friends replied, "PLEASE tell Kimmy that I'm soooo glad she and her friend are okay. This could be a very different day, but it's not!!! Her guardian angel was on the job." And "A+ for your guardian angel." We wanted to believe my late mom had watched over me. This bothered me somehow. Too easy, too neat. What about those who died? Didn't they deserve a guardian angel? Do their angels get an F? Assuming mine was on duty, what an amazing job she did. Where was she located? Could she have swum, loose and free, in the splash that rose up out of the water? Could she have throbbed between the spaces of people's broken bones, like for one survivor named Hector, who was thrown to the ground? Could she have rocked beneath the water like a great giant hand, catching the steel as the bridge crashed? Might she have soothed any of the thirteen who died as water suffocated them, windows refused to unroll, and doors refused to open? Did she zoom with other angels, invisibly flitting through ruins to quickly ease pain? Might she have surrounded a woman named Sadiya, her daughter, Hana, and unborn baby, pulsing them with love?

After work I drove to downtown Minneapolis to pick up Rachel. To the left a neon sign for Murray's restaurant glowed: HOME OF SILVER BUTTER KNIFE STEAK. All along the sidewalk, people sat at tables for happy hour. To the right City Center. Ahead the Hennepin County Government Center building, big as a city block. Parked on Sixth Street, the wind blew so hard, it jostled my four-door red Toyota Prius. The wind in the background of the radio whistled: I imagined a tornado sounded like this as it would build. The car shook and trembled. My nerves came undone. My heart beat faster. The car kept shaking. A few times it almost lifted. I watched a black sedan in front of me back up. When would they stop? When they left the curb, I surveyed other cars on the street, checking for movement. Dorothy from *The Wizard of Oz* popped into my head. I thought of tornadoes last summer in Hugo, Minnesota.

"Please don't let my car lift up or fall down," I said to nobody.

My brain presented images unbidden, and I sorted them. In this way my brain worked overtime. I grabbed a book from the door pocket, but one page in, I stopped. Don't lose yourself. Pay attention. I looked out at passing cars and wondered if paying attention can actually keep bad things from happening. I realized that was silly and returned to my book. Then I stopped again. I couldn't sense myself in space. The street vibrated as trucks and buses lumbered by, and I fought visions of everything falling into an abyss. I saw the road open like in New York, July 2007, when an eighty-three-year-old steam pipe (laid in 1924) exploded and blew a hole in a city street, swallowing a truck, killing one person and injuring others.

10

For the 35W Bridge collapse, was there proof that someone had knowledge? Did a smoking gun exist—or survive—evidence of a person or persons who looked the other way and can be named? A trail and a puzzle lived within the inspection reports. Bridges have their own terminology. Could I piece together enough of this overwhelming flood of information to understand what happened? I didn't know. Sometimes I thought it impossible, but over the next decade I would try.

11

Basic Bridge Terminology

Abutment: Refers to a place where two structures meet—basically a retaining wall designed to carry the loading conditions present in bridge structures (piers connect to them). For example: the structures at the ends of a bridge that join the bridge slab to the roadway.

Deck: The part of a bridge that directly carries vehicular and pedestrian traffic.

Substructure: The parts of a bridge that distribute loads to below-ground footings. Pilings, shafts, spread footings, and columns may be part of the substructure.

Superstructure: The parts of a bridge that connect the deck to the substructure—includes girders and bridge railings.

12

Inspectors nicknamed Federal Bridge 9340, the old 35W Bridge, the "Squirt" Bridge. It's unclear how it got that nickname. The steel deck truss was nonredundant, which means if one part failed, there's no backup and the entire structure could tumble. A truss exists anytime timbers or steel join together forming triangles: a roof is a common truss. The latticed steel beneath the bridge deck was composed of trusses. Built in 1964 by

Hurcon Inc. and Industrial Construction Company, the steel trusses and deck constructed by Industrial Construction Company in the summer of 1965, the bridge opened to traffic in 1967. It was 1,907 feet long, had three lanes of traffic north and south, and around 140,000 cars used the bridge daily. Before it collapsed in 2007, it was young for a bridge, open just forty years; it rated a 40 on the 100-point National Bridge Inventory scale: structurally deficient.

13

On Monday after the collapse, August 6, I return to the Carlson Towers, to my job as a technical writer.

It's my first day back, but I can't concentrate on anything. A lot of people want to hear the story. I hate being away from Rachel. It's starting to sink in—the potential to be ripped from each other. It hasn't even been two years since our Canadian nuptials. Our nuptials at home were just last summer. I really hate being away from her. We are clinging to each other more than ever.

So much that we appreciate all the small things. Like my lunch—on top of bread, lettuce, tomato, and turkey bacon, written in Rachel's hand, three simple words on a white square of paper: "You're my favorite."

I looked the same, but I felt different. Someone dropped a ream of paper, and I jumped through the ceiling. It took a long time to calm down. I picked a scab on my back, and it bled. I felt a jumpy energy. Standing in the basement cafeteria, I looked warily at the drop ceiling. What if it fell? You must be on the lookout. Anticipate danger, prevent yourself from being killed. If only we could say, "Nah, don't worry about it. It could never happen." Logic reasons, now I should fear everything.

My arm hurts from the tetanus shot I got at the emergency room. Right now this seems to hurt more than my neck, which is subluxated (bones are misaligned). Emotionally, I'm not sure. My short-term memory is bad, and I feel tense and vigilant everywhere I go. I try to relax, but it seems out of my control. A counselor tells me that the bad short-term memory is a common symptom that people experience after trauma. A friend tells me,

"I can understand you not wanting to stay at work and being distracted. It will take some time to get back in the groove of everything."

The parking ramp at work scares the hell out of me. While I was walking, a car drove over the seam, and the floor shook. I froze. I clutched my heart and stood still, seeing the whole thing falling. The concrete all around looks too heavy and too big. I am too small, walking on these ramp floors and several stories above me. I think, Oh god, please don't let it collapse. I realize that if it does, this time I won't have a chance.

14

Federal Highway Administration (FHWA) Condition Rating Scale

The Federal Highway Administration uses a 10-point rating scale as one method of categorizing the condition of bridges. This scale is used in bridge inspection reports and is useful information when looking specifically at the 35W Bridge.

9 Excellent
8 Very Good
7 Good—No problems noted
6 Satisfactory—Some minor problems
5 Fair—All primary structural elements are sound but may have minor section loss, cracking, spalling, or scour
4 Poor—Advanced section loss, deterioration, spalling, or scour
3 Serious—Loss of section, deterioration, spalling, or scour have seriously affected the primary structural components. Local failures are possible. Fatigue cracks in steel or shear cracks in concrete may be present.
2 Critical—Advanced deterioration of primary structural elements. Fatigue cracks in steel or shear cracks in concrete may be present or scour may have removed substructure support. Unless closely monitored, it may be necessary to close the bridge until corrective action is taken.

1   Imminent Failure—Major deterioration or section loss present in critical structural components, or obvious loss present in critical structural components, or obvious vertical or horizontal movement affecting structural stability. Bridge is closed to traffic, but corrective action may put back in light service.
0   Failed—Out of service; beyond corrective action.

Think of 1s and 0s as the rotten apple scores, for "imminent failure" and "failed," past tense. Bridges that rate 0 to 4 are considered "structurally deficient." The FHWA insists that this rating does not inherently mean that a bridge is unsafe. The old 35W Bridge rated a 4. A "functionally obsolete" bridge only means that the design is outdated—that the structure may carry a greater load than was originally intended.

15

FHWA Sufficiency Rating Scale

Sufficiency Ratings, another method for measuring the condition of bridges, are on a 0-100 scale; used to determine eligibility for federal funding.

16

Basic Bridge Terminology

Girder: A main support member for the structure that usually receives loads from floor beams and stringers; also, any large beam, especially if built up.
Pier: The main vertical supports that hold up the bridge deck, at an intermediate location between its abutments.
Reinforced concrete: Concrete with steel reinforcing bars (rebar) bonded within it to supply increased tensile strength and durability.
Steel: A very hard and strong alloy of iron and carbon.
Tension: A force that pulls or stretches.
Truss: A truss exists anytime timbers or steel join together forming triangles, so as to support longer spans. A roof is a common truss.

17

Special Inspection: August 27, 1971

Inspection reports for Bridge 9340—35W Bridge ("Interstate 35W Bridge: Original Plans & Details"). Special inspection—typed report: "Evidently this is first report since traffic has been put on." Ratings of 9, new condition, and 8, good condition—no repair necessary. Notes for item 18, area under bridge: "Must be 200 wine bottles and clothing, bedding etc. under So. Abut. (Hippies Haven)." Estimated remaining life: fifty years (2021).

18

I pull open the shower door, step over the edge onto the white ridged floor, pull the lever, and the water pours out. Surrounded by three gray tile walls, a glass door, and a tile ceiling in a small space, I start most days with the sound of water.

Stand to the side. Wait for the right temperature. My gaze fixes past the scrubby hanging on the wall. Warm now, I enter the stream and soak my hair and skin and wash away the day and the night with my head tipped back. My hands rinse time and memories. I shampoo my hair into a lather, working soap through—this water is safe. It pours from the showerhead in an expected way, not suffocating, dirty, and polluted like the Mississippi.

Walking into the shower the morning after the collapse, water set my skin to burn as cuts and bruises got wet. Every day I should shower and feel peaceful, I think. I didn't die then. Could've. Maybe should've. Something kept me just above that river. Nothing seems the same. Water on my skin, no pain now. I remember the thirteen who died; remember my panic when the river poured into the car and water splashed in the seams: when I exhaled my chest to a concave shape, my arms flailing in a "what should I do?" gesture, seatbelt still on. Quick, get out of here!

Pick up the shaver. Roll over calves. *Why'd that bridge collapse have to happen? How could they have let it happen? Doesn't matter what the State of Minnesota, MN/DOT, or the NTSB say. Those thirteen won't come home. I could've been one of them. I could've been the fourteenth. I can never forget this.* Shampoo suds slither around me and disappear down the drain. Stopping at the drain, at a bottleneck, water slowly breaks up the foam. Every day repeats this: I step into the shower, under a rain cloud. Step out. Thank god I lived.

19

Dozens of people survived the I-35W bridge collapse. Some were seriously injured, others escaped with minor injuries. But all are experiencing the emotions that come after surviving a brush with death. Experts say understanding those emotions will take time and patience. St. Paul, Minn.—Kelly Kahle and Kimberly Brown were crossing the I-35W bridge on their way to a soccer game when the bridge fell from under the car. . . . Brown says, "It was the most terrifying thing I've ever gone through. I thought we were going to die. The fact that we're living is just as much [of] a shock." . . . She says she's talking about the experience a lot, but she's not sure how to get past the fear and anxiety she feels. "I just feel so lucky to be alive," Brown says. "There's definitely part of me that's questioning why was I spared. What am I supposed to be doing with my life?"

Dan Gunderson, "Two Survivors Were on Their Way to a Soccer Game," Minnesota Public Radio, August 6, 2007

On the one-week anniversary of the bridge collapse, Minneapolis National Night Out gatherings observed a minute of silence for the bridge collapse victims, their families, and emergency responders. A minute later City Hall and churches rang their bells. Minnesota Public Radio aired a remembrance special. Kelly and I had interviewed with a reporter named Dan Gunderson, and our talk aired about fifteen minutes into the show.

My cuts were healing.

The National Transportation Safety Board (NTSB) started a toll-free number and asked survivors to call with their accounts. They had no record of me, probably because I was a passenger and not a driver. They said they were backed up and might contact me later.

I contacted the Red Cross to ask if a support group had started. "It's too early," they said. Too early? I needed to talk to people now. Instead, the Red Cross connected me with one of their volunteers, whom I met in downtown St. Paul. We tromped through a musty building looking for a conference room. Then we sat in a windowless room at an oval office table with plastic chairs. I felt no rapport with the counselor, so I didn't go back. Instead, I went to work with a gloomy feeling that life was all totally pointless.

Away from home the world felt compromised, full of sharp edges, rickety stairs, double-digit floors that could fall from the weight of too many people, computers, and soulless commerce. Any moment buildings could collapse.

With the bridge out, Twin Cities' drivers had to navigate around the collapse site. The Minnesota Department of Transportation (MN/DOT) posted hunter orange signs. Block letters warned, ahead the "freeway ends." How laughable. More like, the whole stupid road was broken.

20

Basic Bridge Terminology

Beam: A linear structural member designed to span from one support to another.
Stringer: A beam aligned with the length of a span that supports the deck.

21

Inspection Report: December 27, 1973

Annual inspection—typed report: "Traffic has been opened to entire structure since last report. Cracks have multiplied considerably and laminated areas are numerous—recommend cracks be sealed. Condition Ratings: Substructure 7, Superstructure 8, Deck 7. Estimated remaining life: 43 years [2016]."

22

Post-collapse brings dreams of the fall. Terror follows me into sleep. Upon waking, my jaw aches, and I realize I have been unconsciously clenching my teeth. *Tight. Tighter.* I crack a tooth, and I'm on my way toward breaking the rest of them. My teeth break, starting a slow process over time—days, weeks, months, a year. I have no idea how to tell my sleeping mind to relax.

I didn't want to go to sleep at night, so I stayed awake as long as I could. I surfed the Internet looking for stories recapping the bridge collapse. "Why?" I asked myself at 2:00 and 3:00 a.m., eyes bleary. I was there. Why would I need more proof that this horrible thing happened? When I succumbed to sleep, exhausted, later I'd wake, back aching, and mourn the losses happening to so many innocent people. Massive injuries, broken bones, people robbed of a proper good-bye. Our room's royal blue walls lulled me in my reverie. With no preparation against the attack of grief, Rachel held me, cried with me, and rocked me as I sobbed, her hand resting lightly on my back. She handed me tissues. The thin, gossamer material disintegrated under the tears falling into my clutched palms.

Lucy stood from her dog bed on the floor and set her muzzle at my side. Rachel and I looked at each other through stinging, swollen eyes. All cried out, we hugged, kissed, and said, "I love you"; "I'm so lucky to have you"; "I'm so lucky to still be here"; "What if I had died?" "How would I have gone on?" "Those poor people"; "I was only ten seconds from being crushed by that sign"; "I could've been five seconds faster and drowned"; "A woman stood on a car's rooftop but couldn't get to that other woman as she drowned"; "Thank god you survived. Thank god."

Then awash with grief again, guilt coated me in the darkness like paste. Our momentary calm dissipated into more crying, with the thick realization of how lucky we were. This was my experience of survivor guilt—the persistent suspicion a survivor can feel: that they didn't deserve to survive and therefore don't deserve to be joyful. Can't even temporarily forget those who were lost, those who suffered and grieved.

Lucy wiggled at the bedside, paced, and licked her chops, which made a dry sticking sound.

Trying again to sleep, my left nostril closed. Awareness shifted to my heartbeat. Thump. It beat in my ears. Thump. Through my nose as I inhaled. In pitch black, was anything real? Eyes half-open, I rose and hobbled, barefoot, into the ecotone between chill hardwood and then tile. Gradually, I woke, the haze lit by nightlight. The clock ticked lethargically. Time couldn't be this slow. Batteries must be weakening. My mind turned on repetitious thoughts. This is real. You're still alive. I wasn't just afraid of The End, the moment when living became an impossibility, senses shut down, and the microcosm of the world became a narrowing of light—to not wake again but also to know reality will vanish. The human condition. Tears on the pillow won't change it. Repeatedly, I realized a psychic truth. We're afraid of being lost, of suddenly being gone. After pupils adjusted, I saw by moonlight, which illuminated the cloth blinds over the rectangular windows. In the semi-asleep state of the house, I kissed Rachel's cheek as she slept. She stirred and mumbled, "Thank you, baby," and changed positions. I closed my eyes, rooted in that room to the wind from the sound machine. Still alive.

Planes roared as they passed over our house at night. I tensed, paused, listened. Will they fly over successfully? How many tons do they weigh? Will one fall on me, on our house? Reading in bed, would I be found like those people in Pompeii millennia ago, frozen in time? When I should be letting go, PTSD knocked at the door of my security. Lucy sighed and turned onto her back, exposing her belly, splaying her back legs. She wagged her tail and stuck a paw straight into the air. I petted her and listened as we lay in the flight pattern.

I tried to relax, but it seemed out of my control. Short-term memory was nonexistent; I seemingly couldn't remember anything. It felt ridiculous. Apparently, this is common after trauma. I forgot whole conversations. At times I had to stop people and fess up. "I'm not here, sorry. What were you saying?" Floors shook in ordinary places—a friend's upper-level porch, a rooftop bar.

23

Basic Bridge Terminology

Fatigue: Cause of structural deficiencies, usually due to repetitive loading over time.

24

Inspection Report: June 27, 1974

Annual inspection—handwritten report (difficult to read): Condition ratings are: substructure 6, superstructure 8, deck 7. Approximately fifteen-foot-long vertical crack on river side of pier 7 (west) column; a smaller crack on opposite side. "Estimated remaining life: not noted."

25

August 7, Tuesday. First visit to chiropractor. We had Skinny Cow mint ice cream sandwiches in the freezer, and I just ate the last one. I decided to soak the plastic tray before recycling it since there are bits of green ice cream and dark brown sandwich stuck to the sides. I set the plastic tray in the kitchen sink, open-faced, and turned on the faucet. Chunks of green ice cream loosened, floated, and poured over the edge.

Look at that. I watched the bridge collapse reenacted in miniature.

My thoughts seemed to always be on the bridge. It's a big mystery. *Why I was there? Why I was spared? Why were other people hurt so badly? What would have happened if I had been five minutes late? Or early? It's not fair, especially to those who died. My cuts are already almost healed.*

One minute I thought I could comprehend it, and then the next I remembered falling and the incredible sound of crashing. And the terror that I felt, with each passing moment, that I might be crushed came back in a stark and awful way.

My life seemed to be humming along before the collapse, but afterward

I felt confused. I questioned everything I used to do. I wondered, "What am I doing here?" Was the universe trying to send me a message? The poet Mary Oliver said, "What will you do with your one wild and precious life?" I felt inspired to do more, but what?

26

Basic Bridge Terminology

Corrosion: The gradual destruction of material, usually metals, by chemical reaction with the environment.

27

Inspection Report: June 27, 1974; December 22, 1975; and June 28, 1976

Annual inspections—handwritten report: Condition ratings are: substructure 6s, superstructure 8s, deck 7s. Inspector repeats comments from 1973 on each report. Cracks, expansion joints, flanges. Reports from this point forward: "Estimated remaining life: not noted."

28

The day after the one-week anniversary, Rachel had a night class at Metro State University, but I wished she didn't have to go. She didn't want to be away from me either, but she couldn't miss, so we compromised, and I rode along. That way we'd have the trip there plus the trip back: an hour of extra time. While she was at class, I'd find somewhere to write and catch up on calls. This was separation anxiety, but we didn't care.

As Rachel drove, I stared out the car window and watched the passing lines of trees, fluctuating canopies of green.

I repeated one victim's name.

Sadiya Sahal. Sadiya Sahal. Sadiya Sahal. My broken big toe ached. The aching in my body started my thoughts about the people thrown into the river, unable to escape.

What was drowning like? Horrible, I heard myself answer.

I imagined my throat full of water. I held my breath until it burned.

When a person loses consciousness, when did it end? Any pain I felt, should I welcome it? All the while I said nothing. Then a driver honked and snapped me from reverie. My heart pounded as if against a cage.

Blunt force injuries. My mind repeated that phrase. Blunt: a dull unwieldy object. Force: an object in motion stayed in motion until something of equal action stopped it.

Injuries: breakage, severance.

29

Basic Bridge Terminology

Spall: When material chips, crumbles, or flakes from a metal.

30

Inspection Report: May 24, 1977

Annual inspection—handwritten report: Comments repeated from last inspection reports. Cracks, expansion joints, flanges. "Estimated remaining life: not noted."

31

I noticed weird thoughts. I wondered but didn't judge. At Paradise carwash attendants towel dried my car. When they finished, it gleamed. I handed them my claim check and tip, got in, adjusted the seat they'd pushed way back, and then drove away, admiring the crystal view in the rearview mirror, enjoying the lingering smell of window cleaner in the cabin.

Three days later my car still shone. This may be a record. But then I imagined it being totaled like the ones on the bridge. When you fall from eleven stories up, did any of these washes matter? If you had known, think of all the things you could have skipped. You could have saved your money instead of paying for that last oil change. Put off getting those new tires.

One day, commuting home, I drove in perfect weather: sunny, low humidity, windows down, a steady breeze. Stop, go. One car length. Stop again. I started feeling bored. On the opposite side cars zipped past. Cars and trucks looked different, like matchbox models: impermanent and crushable. I wasn't on a bridge.

My mind wandered. What if a car skipped the barrier, flipped hood over trunk in an aerial stunt? An ambulance rushed on the other side, siren wailing. Behind my eyes, burning.

Another morning my body tensed. Cars pulled to the side. An ambulance screamed past and then disappeared. I started heaving. I blinked rapidly to overcome tears, then gripped the steering wheel tight. Brake lights illuminated in a long line ahead of me. Acutely aware of each moment. When the bridge collapsed, it all changed instantly. Should I ever forget that?

32

Basic Bridge Terminology

Section loss: Loss of original metal.

33

Inspection Reports: 1978–80

No annual inspection report for 1978, 1979, or 1980.

34

Two weeks after the collapse, I would board a plane to New Jersey, a family trip planned long ago. I wasn't looking forward to the traveling part at all. I've always been uncomfortable flying, but now it was much worse. Images of the collapse were fresh in my mind, the windshield turning brown as we fell—concrete and pieces flying, the sounds of crashing, the car plummeting. For the first time I was genuinely concerned my body might forsake me. I could see it. The plane would lift off and shake, engines roaring. I'd fall prostrate in the aisle, wailing. My mind constantly saw images of the collapse. What if I lost control, actually lost my mind? I decided, no fooling around: get meds. As I waited on the phone for the nurse, I thought, "Oh, fuck. I have to get on a plane," a lump rising in my throat. In my mind I jettisoned above New Jersey's bridges, many of them also rated fracture critical. Many of them rickety, old, and tall. I told the nurse what happened. Could I hold? Yes. Then she came back. "Oh my gosh," she said, "No problem." They prescribed Lorazepam (generic Ativan; belongs to the drug class benzodiazepine, which act on the brain and nerves to produce a calming effect). Yay, drugs.

The day arrived. August 16, and we boarded the flight to New Jersey. I wanted to blink my eyes and magically teleport to Ocean City. Sitting in coach, I expanded the seatbelt and turned toward Rachel, balled up my jacket, and buried my forehead against her shoulder. So drowsy the meds made me. My hand in hers, I shut my eyes to a restless sleep and placed faith in her warm palm against mine. I noticed the roar of the engines, the vibrating wings at takeoff, shifting movement, regular bumps and the occasional drop. But my mind disconnected the dots. My body failed to go into fight-or-flight mode. Instead, deadened recognition. Apparently, in the event of a crash, I would observe and not participate.

The drugs worked reasonably well, like a marvelous and mysterious friend.

When we made it, I toured my sister's summerhouse with my family. I followed quietly, overcome with gratefulness. I pulled out a patio chair and sat on their third-floor balcony and listened to insects and ocean waves. Below, landscapers laid an intricate brick path. Lush grass filled S-shaped borders that lined bushes and trees like a painting. Between the house and the sea the terrain changed from tall grasses and sand dunes to taller rougher grasses, then more sand. People speckled the beach, and a lifeguard on a Jet Ski patrolled near shore. A steady breeze tickled my chin.

Later, walking along the beach, I stepped to the sounds of constancy. Water, its rhythmic undulations, pounded with waves of persistence. Wading, the water rushed in at mid-leg. I stooped and lifted shells. Most in pieces, I scraped at their rough edges, stroked the grit, and mulled over being alive.

One of my nephews, Billy, just turned fifteen, stood lanky and already taller than my five feet one and a half. Boogie boarding, he ran along the sandy shore, fell off, balled over giggling, then glided back on. One of my nieces, Alexis, twelve, also taller than me, scooped handfuls of sand. Natural curls disguised the "Wow" that escaped her mouth as she ambled over.

Showing me a small shell, she said, "Look, it'll move."

"No way," I said, and looked closer. When it did, I said, "Eew!" and recoiled. "Go show Rachel!"

She wetted it and leaped away. From a distance I saw Rachel's mouth twist. Smelling the salty air, I felt small on that shoreline, miles of ocean outside of me.

My biggest decision that night: did I want a hot dog or a hamburger? Family bustled about. Within the hum of activity, I felt insanely happy. John Mellencamp crooned from my brother's iPod. The words sounded new. *Hold tight.* Everything felt like a metaphor. *Don't even know if I'm doing it right.* I didn't know either. I just lived on.

I cozied up to my dad on the deck. At first we didn't talk. Near the railing, the song rung clear through the screen door. Hunched from his old football injuries, he wasn't much taller than me.

Dad—Bill "Boom Boom" Brown, "Old #30"—had been a star running back in the National Football League for fourteen years: thirteen seasons with the Minnesota Vikings (1962–74). He held several Vikings records: all-time leading rusher (1,627 yards), tied for third in touchdowns (52), and fourth in career points scored (456). Muscular and sturdy, five feet eleven, bowed legs, sideburns, and a signature flat top, he was a cog in the Coach Bud Grant machine: a legendary squad of players comprised of Fran Tarkenton, Grady Alderman, Mick Tingelhoff, Chuck Foreman, and Alan Page. In NFL Films highlight reels narrators described his trademark running style, how he ran through opponents instead of around them, at the open-air Met Stadium—where the unpredictable and harsh Minnesota weather was as much a part of the game as the gridiron, the ball, and the goalposts. Revered for his toughness, he played in blizzard conditions in short sleeves. At the sidelines, his breath visible like fog, he called out to players on the field, "Hurts a little, doesn't it?" In those days, when the yellow goalpost stood on the zero yard line (instead of at a distance, like it is today), he was knocked unconscious when he hit his head on the goalpost when running in a touchdown. Carried off the field by stretcher, he was dubbed "Boom Boom." A nickname that stuck. In later years he will battle dementia and CTE—chronic traumatic encephalopathy—brain injuries that are just now being addressed in today's NFL.

Growing up, I had always been closer to my mom, but at that moment it felt natural to seek physical closeness with my dad, which wasn't a usual thing. I took his arm, and it felt surprisingly normal to put it around my neck. We stood silently. What had he been thinking? I thought about the colossal earth. How at the New Jersey shore or at home in Minneapolis, where the big bridge fell, I was lucky enough to continue. "I miss Mom," I said, telling him my thoughts, which felt unfamiliar, but I did it anyway. "I wonder, was she my guardian angel like others have said?"

"I don't know, honey. She did a helluva job if she was."

Blackness. Waves massaged the beach, the wind whooshed, the sounds of the boardwalk bustled, but the lights and the white dotted carousel had been turned off. I felt the air but saw nothing. After being alive was done, "in death," could that existence (if it's an existence at all) be like this? Looking into a charcoal night, knowing something is there, trusting what can't be seen. That instant eleven had died, and divers searched for the last two victims who remained missing beneath that dirty smelly water—below twisted steel beams, girders, and pavement. Minneapolis officials had switched from rescue to recovery. The Army Corps of Engineers adjusted the lock and dam to aid the recovery effort. With the river lower, I finally saw how Kelly's car stayed up, broken concrete below the rear bumper, precariously straddling a gash in the deck that had opened just enough. Any larger, we might have drowned, tangled in the trusses. It doesn't matter that I've flown hundreds of miles from home. The facts remained. Thirteen perished.

35

Basic Bridge Terminology

Scour: The removal of sand or rocks from around bridge piers or abutments (e.g., by force of water or flood).

36

Annual Inspection Report of Bridge 9340 (35W): October 18, 1993

Bridge inspectors specifically state that they saw a gusset plate (in fact, one of the plates that fractured in the bridge collapse) with a "loss of section 18″ long and up to 3⁄16″ deep (ORIGINAL THICKNESS = ½″)." The state knows how thin the plates are, over a decade before the bridge collapses. They even have photos of the gusset plates bowing. Here's just one photographic example from June 2003. With only the naked eye, we can see the steel *bending*.

5. Bent U10 West gusset plate. Photo by National Transportation Safety Board.

In August 2008, when the NTSB comes out with its final conclusion (a year before the bridge collapse), the safety board will profess that the thickness was never known, and we'll all believe it because we're not bridge inspectors. But it was known. When the NTSB also holds a preliminary press conference even earlier, in January 2008 (less than six months before the bridge will collapse), to announce that the thinness of the gusset plates wasn't known, we'll all believe that too because *its inspectors are the authority* and we're relying on them. This will be critically important later because what the NTSB has told the public—blaming the collapse on a design flaw—doesn't jibe with overwhelming evidence and what professionals in the field have concluded. The bridge collapse unequivocally, absolutely, could've been prevented.

```
·'Bridge No.:   9340      Bridge Inspection Report      Oct 18, 1993    Sheet 5 of 6
           LOOSE AND LEAK.

WESTSIDE:

LOOSE BOLT 2ND INTERIOR STRINGER BEARING AT V18

NICK ON BOTTOM OF DIAGONAL L15 - 14

NICK ON BOTTOM OF LOWER CORD L15 - 14

2 NICKS IN DIAGONAL L15 - V12

NICK IN BOTTOM OF TOP CORD L10 - V8

NICK IN BOTTOM OF H SECTION TOP OF FLOOR BEAM V6

NICK IN TOP OF H SECTION BOTTOM FLOOR BEAM V6
------------------------------------------------------------
 ADDITIONAL COMMENTS FROM OCTOBER 13-18. 1993 SNOOPER INSPECTION.
20
  #) DOWNSTREAM TRUSS AT L11 INSIDE GUSSET PLATE HAS LOSS OF SECTION
     18" LONG AND UP TO 3/16" DEEP (ORIGINAL THICKNESS = 1/2").

     DOWNSTREAM TRUSS AT L13 THE LOWER HORIZ. BRACE BETWEEN THE TRUSSES
     HAS 3/16" SECTION LOSS AT RIVITED ANGLE.

     TOP CORD OF UPSTREAM TRUSS JUST NO. OF NORTH RIVER PIER - POSSIBLE
     CRACKS IN WELD OF WEST BAFFLE GUSSET TO TOP FLANGE.  CAN'T GET TO
     IT.  CHECK AT NEXT IN DEPTH INSPECTION, POSSIBLE ULTRA-SONIC
     INSPECTION.

33#) AT FLOOR TRUSE #11 AT STRINGER #11 THERE IS A CRACK IN THE WELD
     FROM THE BEARING BLOCK TO THE TOP FLANGE OF THE FLOOR TRUSS.

     AT FLOOR BEAM U7 UPSTREAM SIDE DIAGONAL TO THE NORTH HAS EXCESSIVE
     PLAY & MOVEMENT AT UPPER PIN - PIN SHOWS LIGHT WEAR, 1/8" GAP.

     FLOOR TRUSS #1 - COTTER PIN MISSING ON PIN HOLDING SWAY BRACE TO
     LOWER CHORD.
```

6. Bridge 9340, October 18, 1993, inspection report, sheet 5 of 6. Minnesota Department of Transportation.

37

"Construction Expert Denounces the NTSB's Report on the I-35W Bridge Collapse"

In November 2008 an article was published by construction expert Barry LePatner, author (with Timothy Jacobson and Robert E. Wright) of *Broken Buildings, Busted Budgets: How to Fix America's Trillion-Dollar Construction Industry*, blasting the NTSB report. "One important factor contributing to the poor state of America's infrastructure," LePatner wrote, "is the seeming irresponsibility and inefficiency exhibited by those who have been elected or appointed to government positions that supposedly exist to ensure the safety of the public."

To illustrate, LePatner identified several red flags that should have warned MN/DOT and other officials that the I-35W Bridge was in trouble but instead were ignored, misunderstood, or simply not acted upon in time:

The I-35W bridge was first rated in inspection reports as "structurally deficient" in 1990. Despite annual reports describing a continuing section loss and build up of corrosion at key places, as well as the attention of a number of consultants who recommended substantial remedial action be taken, at no time between 1990 and its collapse in 2007 was the I-35W bridge's condition ever raised above its "poor" rating.

Photographs exist of gusset plates "bowing and arcing" as early as 2003, but the photos, taken by MnDOT consultants, were apparently dropped into a file folder and forgotten. MnDOT inspecting engineers did not deem these red flags to be serious enough to command attention.

In 1996 a bridge on I-90 outside of Cleveland with a structure similar to the I-35W bridge partially collapsed as a result of improper gusset plate design. But although a) Federal officials investigated this serious failure, b) an official report from outside engineers was filed indicating that the gusset plates did indeed contribute to the bridge's collapse, and c) *Civil Engineering*, ASCE's monthly magazine, published an article in 1997 detailing the Ohio bridge collapse, officials at MnDOT denied ever having heard of the Ohio bridge failure and said they were unaware of any prior problems with gusset plate design.

Discussions concerning the need to add redundancy to the I-35W bridge had been underway years earlier—but action was never taken. And, in fact, MnDOT instead scheduled redecking work that overloaded sections of the bridge, and, according to the NTSB, contributed to the eventual failure of the gusset plates.

38

Hands—numb. Clench, open. Tingling. Ache in neck. Resistance. If I could only bend it far enough to crack it. But where, which way? Rotate and stretch. Oh. Oh. Something is off. Pressure with the slightest change in neck position. My head is like a bobblehead doll. I self-correct countless times throughout a day, my head too far forward. I push my chin under and back. My neck has subluxated again—"gone out," as it's informally called. Pain and numbness radiate into my hands and my collarbone.

Back home in Minneapolis, bruises and cuts healed, I had injuries no one can see. Results from X-rays done before Jersey were in. X-rays revealed spinal compression and, from top to bottom, a slant 0.16 centimeters to the left. Instead of a normal sixty-degree curve (imagine an approximation of the curve in the letter *C*), X-rays showed a three-degree curve in my neck, more like the ultra-flat letter *I*. In the future this curve would become one-degree reverse of zero. I would later learn that my new neck condition was called kyphosis, or "military neck."

Remembering the collapse, I could imagine how that happened. Crouched and bent at the waist during the collapse, my head had bounced like a bowling ball. I could be standing on solid ground and feel nauseating waves of dizziness exacerbated by my super straight neck.

Ironic. The obvious metaphor that my spine and neck are my body's bridge. Injured, permanently. Not fixable, only managed. Requires constant and regular care. Sometimes increasing frequency. My neck and spine feel brittle, arthritic. People will see my structurally deficient spine as my fault.

I didn't neglect my body's bridge. It was hurt. I barely got out with my life. That's just the truth. How in the world will I get by when I'm fifty, sixty?

Go in immediately. Get help. Suffering is pointless. If you can alleviate the pain, do it.

Between chiropractic appointments I did home therapy with something called a posture pump. The goal, to re-induce a curve into my neck. This contraption was ring shaped: a round frame made of a hard plastic-like material, with a soft ring of fabric like a halo all along its border. I would place the posture pump on the floor or a flat surface, lie down face up with my head in the middle, and wiggle and adjust until my neck was in the halo just right. After I'd set my neck on this ridge, I used the pump to inflate the ridge. I started with one pump and held for five minutes, then gradually increased. It hurt terribly at first, but this eased with time.

I wanted to see Amy Mattila, a massage therapist I'd seen previously. But I was waiting for all of my cuts to heal. I emailed her about coming in.

"Yes," she wrote, "we can do a relaxing session and some work with letting go of any feelings so they don't get stuck in your body, so to speak. Your body will let us know what it wants and needs."

Years earlier Rachel had surprised me with a birthday gift certificate. Not only was Amy a lucky find and especially skilled; she had other gifts. Perhaps the strongest being an intuitive sense that could almost be described as psychic. Rachel said that Amy knew things about her based on what her body "told her." Based on where her body had pain, where it held tension, where it was ticklish, and where it felt calm. All were messages. I was amazed by what she sensed about me, without my telling her. But many months had passed before I returned. It was expensive, and if it wasn't "medical," then I viewed it as a luxury. That was before I transitioned my thinking.

Post-traumatic stress took hold. At times the reactions of my body were terrifying and out of control. I felt a strong sense that I should invest in nurturing therapies and work through my body's trauma. Amy taught me that the body has a memory. Situations you go through that cause happiness, sorrow, or fear must be processed by your body. When you ignore them, deny their existence, you sentence your body to years of hard time. I learned that healing wouldn't come by running from the trauma, only by going toward it. This didn't happen overnight.

In session Amy palpated my leg muscles. As her hands ran the length of my calves, I thought of one survivor's six-inch scars on her legs, and my eyes filled with tears. Face down, my sinuses pressed into the cradle, I had to stop getting emotional or I wouldn't be able to breathe. Then my thoughts exploded, making connections. My mother's legs, her neuropathy, no feeling, from her advancing diabetes. As a kid, applying ointment for her, checking her legs each night only to have her lose a leg later, missing her. How lucky I was to have this wealth of feeling in my body. A gift. All of these gifts.

The mental injuries were the hardest to crack. I couldn't concentrate. My mood turned on a dime. I was depressed, irritable, and short-tempered. I filled out a Red Cross intake form, for the Minnesota Helps Bridge Disaster Fund—a fund that was created to gather charitable contributions from the community—but the process took weeks, months even. And if I used funds, would it reduce the funds available for others more seriously injured?

My emotional state manifested physically. Persistent, itchy, burning, swelling hives covered my body. I had to get them under control. Benadryl didn't touch them. En route to doctor appointments, I passed a guy roaming the sidewalks, soliciting members for an environmental group. Errands accomplished, heading back to my car, I passed again.

"You're still here?"

"Yep, still here." I looked both ways to cross the street and made it to my car, evading a sales pitch, but his words echoed. When I arrived at my next appointment, I parked, locked the doors, and hustled. People sat outside a café at tables on the sidewalk. One read a newspaper and jabbed at some food on a plate. Two people visited at another table—one talked, another nodded and took bites: an innocent scene in an unblemished neighborhood. Cars passed. Suddenly the ordinary became extraordinary. I walked by on my own power. Limbs intact. No broken bones.

Still here.

39

Fracture Critical Bridge Inspection Reports

Thirteen fracture critical bridge inspections traced the decay of the 35W Bridge. They began in September 1994 and continued until June 2006, thirteen months before the bridge failed. Each year for thirteen years, broken bolts and a rotated bearing block were noted on inspection reports but never repaired.

Thirteen years of broken bolts.
Thirteen seconds for the bridge to fall.
Thirteen killed.

40

On the local television news, reporters announced that a support group for bridge survivors would begin on Thursday, August 23. Kelly declined going, and I didn't pressure her.

Three weeks after the collapse, I showed up at the St. Paul Police Department, a brick building on the industrial Olive Street. But the date had been misreported, and the group had actually met the day before. Only one other survivor, a woman named Kristin, waited in the white tiled lobby. Two others were there, a reporter and a videographer. Would we talk to them for a minute? Such an innocent question. Kristin and I checked with each other with a look. I felt butterflies. Being interviewed on camera made me nervous, I was the type of person who was more comfortable behind the scenes. But we nodded. "Okay, why not?" we said to each other. We entered a generic-looking conference room. I thought my days before the cameras were finished, having already done gobs of press, but the coming months would prove otherwise. Steve, the videographer, set up lights and a camera on a tripod. Then he came to Kristin and me and, one at a time, attached a microphone to our collars, then asked us to string the black microphone cord under our shirts and hook a battery pack to our jeans. The lights reflected against foil, and it was then that I felt the uncomfortable heat of being watched.

It would be the first in a series of interviews in which I would repeatedly recount my memories of collapse. On that day I was feeling excited and nervous and possibly in shock. With time, years, I would shut down, no longer able to talk about the collapse, no longer able to digest the scene verbally, in brief sound bites, the enormity, the ground to cover, too overwhelming and personally devastating. But for that moment it was fresh, and I had a distinct need to talk it over and work it through.

After the interview Kristin said, "You look so familiar," tapping her fingers on her chin. I shrugged, having no memory of having met her before. "Oh . . . ," she said, and dug in her purse and retrieved her phone. Kristin had snapped a picture of the fallen bridge. At the top the arch of the nearby Tenth Street Bridge framed the 35W Bridge wreckage. In the foreground I scaled the rocky embankment. Kelly held my hand as I stepped in the orthotic boot.

"Incredible," I said.

Kristin had been in the car in front of me. The next day we agreed it was comforting to know someone who understands what these past few weeks have been like. I printed the picture and set it on my desk and wondered, how did I survive that? What does it mean? What do I owe the universe? What can I give?

The next week I returned. A nonprofit organization called "Survivor Resources" ran the support group. On the third floor a window overlooked a small sitting area. Two couches and armchairs were arranged in a circle. Police plaques and framed uniform jackets adorned the walls. Small groups of strangers sat together to talk about the specter of falling. Several survivors wore hard fiberglass back casts. Across the white front of one man's cast were the words BRIDGES SHOULDN'T FALL. Young mothers mixed with bachelors and divorcés, artistic types mixed with hunters and bikers, and the fervently religious mixed with the skeptical. We were every sign of the zodiac, boomers and Gen Xers, all races, a slice of the Midwest, some college educated, some not, slammed together to cope with, for some, the worst moment in their lives.

I was driven and obsessed and constantly problem solving in my head, thinking, What can be done? How can I fix this? As time passed, I thought, "What have I accomplished? Can I do something more?" I sent emails noting impending surgeries, recoveries, and birthdays. Families of those with the most serious physical injuries began to organize and hold fund-raisers. I withdrew five hundred dollars from my savings for various survivor benefits. Why should others bear the burden of astronomical medical costs alone?

Our experiences certainly weren't all positive. At one benefit I introduced Rachel to a fellow survivor.

"Bob, this is my wife, Rachel."

To this, Bob (name changed to protect identity) scoffed and rolled his eyes at the lady friend he had brought along.

During the coming months I signed Caring Bridge websites and attended survivor benefits, the support group, hearings at the Capitol, and lawyer meetings. The events were constant, but each day I lived was extra, so no complaining. Thirteen others didn't get the chance I had to live or to feel sad or angry. I wasn't supposed to have survived. I had to channel a higher purpose. What could I do that the dead could not?

At group we talked about a therapy called "Eye Movement Desensitization and Reprocessing," or EMDR, purported to be effective for PTSD, that would later be crucial in my recovery. I found an EMDR-certified therapist, but she was out of network. I saw this therapist for a brief period. We tried EMDR, but it didn't feel right. Invented in 1987 by Francine Shapiro, it's based on the theory that bilateral stimulation of the brain helps human beings process events. I felt the therapist's eyes on me as her hands patty-caked my knees. I didn't usually mind being touched, but I couldn't get comfortable. At my request we stopped doing it. Without it I failed to get better. For a while I went without counseling because of the cost ($140 per hour). She offered to see me for half that, but over the long run it would be out-of-pocket and still too expensive. I tried an in-network therapist, but it was, to be kind, not a fit.

There are other therapies that exist for dealing with trauma and to try to reduce its devastating effects. Cognitive talk therapy. Virtual reality. One therapy developed in 1997 was a test project, used on soldiers afflicted with crippling PTSD, called "Virtual Vietnam." In 2007 "Virtual Iraq."

I had no hope that I could be fixed. My logical mind was in charge. It said no amount of talking or brainwashing can change the facts. If you're going to die or experience a painful early death, therapy isn't going to do a damn thing. But I was sort of missing the point. EMDR would play a role in my recovery, eventually.

Kelly's reaction to the collapse had been opposite of mine. The day after, she returned to her teaching job like nothing had happened. She also turned to Scriptures, expressing her sense that "thanks be to God" she lived. People react to trauma in discretely individual ways. Some people concluded that since Kelly and I survived together, we must be "best friends for life." Expectations. What causes x does not automatically result in y. A few weeks after the collapse, Kelly and I tried getting to know each other a bit. My big toe mostly healed, we went for a walk. We exchanged Christmas gifts the first year. She gave me a framed picture of close-ups of flowers she'd taken. I recorded food shows for her. One hundred percent comforted by her religion and her church, Kelly wished someday to have a house with a sunroom where, after waking in the morning, she could sit and read Scripture. Of dying she said, "I'm not really afraid—I believe in God."

I didn't know how to explain what I was thinking or feeling. I felt meager satisfaction with the religion that gave her so much comfort. Why had we been together? If it was coincidence, fate, luck, God, or karma, I wasn't going to conclude. Whether or not it worked out the way it was supposed to was unknowable. She and Matt moved out of our neighborhood. They bought their first house, got a dog, and had their first child—a boy they named Josiah, a Hebrew name meaning "the Lord saves." We're rarely in contact these days. We're Facebook friends, and for us it's good enough.

After bonding with other survivors, I learned a crushing detail. Officials found one man who died in the collapse partially in another car, trying to rescue a child. An obscure newspaper story also confirmed eyewitness reports that placed this man on the bridge, walking, after the collapse. He could have been one of us: a survivor. This devastated me. I wasn't in the position to rescue anyone. I thought about his last act on earth, and I felt something like courage build within me. For the first time in my life, perhaps, I wondered if I could be a hero, but how?

41

Minnesota congressman James Oberstar said in an August 6, 2008, newsletter, "We have over 72,000 structurally deficient bridges in the United States—1,156 of them are in Minnesota. It is time to begin systematically repairing and replacing those structures."

In the end the National Transportation Safety Board will blame the collapse on thin gussets, "a design flaw." They will essentially say, How could we have known? We'd never seen the gussets before. This, despite half-inch gussets, eroded from their original size, noted in 1993 and reports ignored year after year. Despite missing bolts, a bearing block turning.

"Federal law requires states to inspect bridges, but it doesn't require states to actually fix them—that's left to the judgment of state bridge engineers and the policymakers who approve money to do the work," according to an article by James Hoppin in a January 2, 2008, *Pioneer Press* newspaper article. "Thirteen years of inspections offer the deepest look into the bridge's condition."

There are 1,097 bridges in Minnesota that are structurally deficient, with sufficiency ratings less than 80 ("Fact Sheet").

Bridges are graded on a scale from 0 to 9. The 35W Bridge received a 4, meaning structurally deficient. The experts tell us this is not necessarily unsafe.

42

A few weeks after the collapse, I started seeing a chiropractor three times a week for my neck and back. I had no concept what that first year would bring. That I'd go to the chiropractor sixty-seven times, that to recover my body I'd exhaust my $20,000 personal injury protection auto insurance. My friend Amy told me this was money well spent. I was glad for her words, but there was this nagging feeling: *It shouldn't have been needed in the first place.*

Coming home from the chiropractor, I thought about the survivors I had met. A woman who couldn't shake the memory of balancing on the roof of a partially submerged car, trying futilely to reach one of the victims. The families waiting while divers searched. The endless wait, losing a quarter of their income, funeral expenses. These people weren't engaging in a dangerous activity. They were doing an ordinary thing, driving on a bridge they thought was sturdy. Ahead the light turned red. In my rearview mirror a woman in a car behind me, squinting into the sun. She could have been someone who died that day, thinking one moment before of dinner or plans. What ordinary thoughts might she have pondered, heading into a brilliant sunset, just before another bridge?

There were other "costs" for those closest to us that are harder to measure. Rachel and I have a strong relationship, but my distractedness, short fuse, constant inner talk, and thoughts about the survivors, the support group, the victims, or all things bridge related were hard on us. Those first months our relationship suffered a dramatic shift. I had experienced a horrifying life-altering trauma, and my wife had not. Whatever her issues, they couldn't hold a candle to mine.

During our everyday discussions I'd reacted harshly, inexplicably angry (a PTSD symptom). We weren't in the same psychological space.

Pre-collapse, whenever we disagreed, we always resolved it before bed. But now I struggled just to function. We had our first fight where one of us decided to leave for the night. After another I slammed our bedroom door and retreated to the couch. Kelley, one of my gray tabby cats, settled between my feet. I needed to weep, so I let myself. The black-and-white cow kitty, Rex, jumped up and climbed over me to join Kelley. They lined up between my feet, pointed the same direction. Flat on my back, I raised my arm and rested it along the back of the sofa and loosely clenched my palm. I thought of a Naomi Shihab Nye poem, "Making a Fist," in which she describes riding in a car, "behind all her questions," feeling sick and wondering for the first time about death, "clenching and opening one small hand." In the dark the haze outlined by the square windowpanes behind my outstretched forearm, skin taut, smooth, still youthful. This night would fade to memory. When the sun rose and filled the room with possibility, I'd start again.

But the blowout fights—when we had to accept that we would stay mad—refused to abate. We'd always depended on the idea of us, of "we." But post-collapse, when we needed each other most, we drifted apart. Rachel would laugh about something—characters on television or some popular gossip—and I would snap, "This is all such a waste of time!" Arguments would start unexpectedly, when we were trying to connect. Rachel is an introvert and is more likely to withdraw, and I'd sense something off-kilter and bug her until she talked. Her hazel eyes reflected, in her downcast expression, the sadness we both felt. People didn't understand, so she bottled up her feelings. She told a friend of hers: "It's hard to compete. Kimberly got the blue ribbon, first place." Inevitably, the discussion would disintegrate, and I'd ooze with anger, saying, "This is not about you." After an atrocious outburst, we had to attend another survivor benefit (we attended many during those months) and pretend we were okay. Rachel sensed me drifting away. She said: "I feel like I'm losing you. You aren't the same person."

She was right. My patience went from high to nil. People's topics of conversation felt obscenely trivial.

One time Rachel was watching *The Real Housewives of New York City*.

"Baby, come watch this," she said, with laughter in her voice.

"How stupid!" I said. "I don't have time for that crap." Contrasted with the heavy stuff going on, I cared not one whit for what I perceived was snarky concocted drama. I wondered, How can people worry about X, when others are in the hospital and can't walk?

I lost my sense of safety. I constantly made comments like "I wonder if a plane will crash into the house." Then I sensed her disapproval and would wish I could take it back. I battled intrusive thoughts . . . had them so often I started keeping them to myself. I got through days with avoidance. I grew paranoid, pervasively on guard, worried that terrible things would always happen. Trying to stop obsessive thoughts was like being told, "Don't think of an elephant." The strain on us was palpable.

Rach told me she missed me. Sometimes I wasn't nice.

There were other ways that the pressure had affected me. We were having family over on a Saturday night, and I got so stressed out getting ready for it. Rachel wanted to cook certain dishes, but I didn't think there was time, and I completely overreacted. I started really weeping. I knew it wasn't about the food. All I could do was just let myself cry.

This will sound odd, but I actually had to work at crying. In some ways I was so strong after the collapse. Deep down, I felt like I was still grieving, still grieving the fact that such bad things happen.

A sober time, not only did I grieve, I brought this feeling to our relationship. Faking it was an impossibility. Rachel sensed my pain and wanted to make it better, but she was powerless and had nowhere to put her own anger and frustration. There were few support outlets for partners of bridge survivors. Rachel wanted me to start seeing someone in therapy and felt it might be good for us to see someone together. Meanwhile, she suffered from vicarious trauma. When you're in a relationship with someone who goes through trauma, because you empathize and try to help, you can also be indirectly traumatized. Rachel looked for someone to talk to, but the therapists at Park Nicollet Clinic wanted to go back into her childhood history.

Rach said, "Jesus Christ, that's not helpful right now. I just need someone to vent to."

She wasn't getting what she needed. I knew it wasn't my fault, but any instability I sensed in her scared me. For the first time in our relationship, I had nothing to give.

Rachel felt like we were going through the stages of grieving for a death, especially the anger and denial. And there was separation anxiety—lots of it. At first even being in the other room, answering emails from friends who were asking about me, sent her into a panic. She'd ask herself, "What am I doing in here, when I could be out there with Kimberly?" Rachel went to school part-time and work full-time, but it didn't feel right. She didn't want to leave and go across town. She felt angry and resentful to have to go out, that there were any expectations on her other than breathing and being with me. Worst of all, she noticed that people who knew what we were to each other would ask, "How's your *friend* doing?" Especially during the first week following the collapse, when things were so scary, hearing me reduced to friend felt like almost losing me again.

Friends and family told me to just enjoy life, and I tried. But instead of feeling at peace, feelings of anxiety worsened. No amount of logic or mental bargaining seemed to make any difference.

Rach said, "You have to get help." I tried to resist this, but it was no use. I started seeing a therapist. This wasn't my first time in therapy. I thought of it differently now. Some see therapy as a means to stigmatize people as weak or signal "they've got a problem," but for me it was like I'd already scaled the sheer rock face. On the other side, I knew, I didn't have to climb without cords, ropes, the right shoes, and a map. En route to the first appointment, I couldn't find it. The office was nearby—I had written down the address and directions—but I started panicking. This was different than just feeling worried; my body seemed to hum like a strummed guitar string. Thinking I was going to miss the appointment, I felt intensely and uncharacteristically angry. Seething internally, I had a grown-up temper tantrum.

Nights when Rachel and I sat on the couch, I said blandly, "I'm just in my own head." In so many words basically I said, I'm here physically, but you know what? I can't deal with your shit right now. But I wanted Rachel's hand to stay, there, on my back, on my body. Another time I said, "I didn't notice when you left the room." Uncharacteristically insensitive, a jerk, I was that absent. My mind worried: it worried on the present; it worried on the past but never the future.

Live for today. No planning. Can't look ahead.

What brought me back to her was simple. Touch. She was always reassuring me. My back, my leg, her hand would rest. It's something I valued and needed, even though I wasn't walking around saying that.

In time I just had to hold her. As much as possible. It was all I wanted to do. Contact. Proof. Home from work, we'd make dinner, feed pets, clean up, then head for the couch. I curled into her side, rested my head on her shoulder, my arm across her chest, a leg across hers, draped like a curtain. We canoodled, carried on whole conversations by touch, held hands and squeezed with Mom's code—one she taught us when we were kids.

I squeezed Rachel's hand three times. I. Love. You.

She squeezed back four times. I. Love. You. Too.

```
            Picture #12
      Missing Bolt @ Stringer #13
       Span #6, P/P-U8, No. B'nd.
```

7. Broken bolt. Bridge 9340, Fracture Critical Bridge Inspection Report, September 28–29, 1994. Minnesota Department of Transportation.

43

First Fracture Critical Bridge Inspection: September 1994

Inspection by Terry Moravec, Kurt Fuhrman, and Pete Wilson. Report prepared by Kurt Fuhrman. Reviewed and edited by Terry Moravec, PE (Professional Engineer).

In panel point U8, pier 6: Bolt broken off, upper floor beam truss and stringer (truss) #13, and the block rotated.

44

For more than a year I avoided the parking ramp at work. When I entered the ramp, I thought, "Here we go." I followed the circuitous route upward to find a spot. I parked four floors up, released my seat belt, gathered my belongings, and exited.

Walking across the concrete floor, the sounds of my dress shoes clicked, and I heard THUD THUD THUD as cars rolled over seams. My body seemed to absorb the sound. My shoulders stiffened, heart raced, and palms sweat.

I quickened my steps, smelled dampness from rain, and felt the THUD THUD THUD of broken bones.

Then a car passed, and the floor bounced.

I froze and involuntarily clutched my fist at my chest, which felt like it could explode. I tried to not crumple. My mind saw the whole ramp implode—all those cars, all the weight, the concrete floor, the support columns—pancaked in on itself, and I had no chance. I would spin into this sinkhole like an egg cracked into a mixer bowl.

Another morning time slowed exponentially as I approached the ramp. Where I should've turned left, I didn't. Coasting downhill with a dull mind, I discovered a surface lot tucked behind the towers in the parking ramp's shadow. Eureka! Why start every day terrified?

I told myself it was irrational to be that afraid, but these were physiological reactions. Attempts to have better, positive thoughts did nothing. I mentally noted the avoidance: classic PTSD.

The rare times I had to park in the ramp because the surface lot was full, I did so reluctantly, all senses alert. Those mornings I searched for a spot near the back, on the ground floor, near the exit. I'd park and start looking. Look at the roof. Look for flaws in the concrete.

I heard thudding overhead as cars passed. (Too aware of sounds: *Hurry. Get out of there.*)

I was deliberate. It was better to avoid walking up the ramp floors. Instead, park at the edges—the floors shake less there—walk over and out, a direct line to the exit, to the surface, to ground.

When I wasn't flashing back, the look of the concrete bothered me. Ramps equaled unknowable pounds of weight over our heads. Or was it tons? One ton equals two thousand pounds.

In ramps my mind performed a kind of math. If a car rolled over my foot, the bones would break. Correct. If a ramp with its concrete floors, supports, ceilings, and hundreds of cars parked high above me collapsed, I could survive. Incorrect.

I couldn't get out of there fast enough. This time I wouldn't have a chance.

45

After the first bolt broke from fatigue, it was replaced with a weaker bolt that broke again and failed to hold the block in place as cyclic vibration caused the bearing block to continue to rotate, or "work," on the one in-place original bolt. This "bearing block" is really a spacer where the weight of the bridge is transferred from the concrete driving surface to "bear," or rest on, the steel substructure of the bridge. Fatigue is failure that happens when the stress varies or "cycles" partially or completely. Fatigue failure is something (but not exactly) like bending a paper clip until it breaks. The paper clip also has something called "work hardening," where the metal changes as it is bent back and forth, and finally it loses the ability to withstand the stress, and it breaks due to fatigue.

46

Walking into the kitchen one morning, I saw the rotting tomatoes. What a shame, I thought, we had just bought them. The once perfect and gorgeous vine-ripened gifts from some farmer's carefully tended field were now rotting. Furry mold and black spots covered the once-taut skin. Resigned, I threw all but three of them in the garbage and thought about all that's been lost.

I headed out the back door. Standing in our backyard with my thoughts, I wondered: *Such a hot day but still in one piece. How is that possible?* Surrounded by a brilliant limerick lawn as compared with broken concrete and rebar, I wondered over my physical presence.

Cold water from the hose ran over my leg, and *flash!* I was no longer just watering dry spots in the late August lawn.

Water fell onto me like hail, the cold numbing my leg. *Rigor mortis.*

Those poor people. Drowning. Blunt force injuries. Was I really here? Still? Could I just be imagining being alive?

This hose might be only a figment of some other reality, I reasoned, when the water covered my leg and chilled my skin. Like the body when it died lost heat. Like the thirteen. Like my mom when she passed. Like ...

My thoughts moved deeper inward. *Oh my god. Why did this happen?* Smell of river rot, water lapping, sirens blaring, and utter quiet when I first set foot on the collapsed bridge, having escaped from the fallen car. I saw the bridge. In my mind it was always there.

47

Second Fracture Critical Bridge Inspection: October 1995

Five main recommendations, two as follows: (1) Several of the bolts connecting the stringers to the floor beam truss (Main Truss Spans) are missing or broken and need to be replaced. NBL (northbound lane): Panel point #8—3 bolts. SBL: Panel point #6—1 bolt, panel point #8—1 bolt, panel point #10—1 bolt, panel point #11—3 bolts. (2) The bolts at the north "crossbeam" connection to beam #3 (span #9, SBL) are "working" and should be replaced.

48

A lot happened in September. Family advised I get a lawyer. I shook my head and said no. I didn't see myself as someone who follows this kind of path. I wanted to let bygones be bygones. But the longer I learned of other people's troubles, the more I realized that this situation was bigger than I'd realized.

One day, parked in the surface lot at work overlooking the five-level parking ramp, I spoke with a paralegal named Lisa Weyrauch from the law firm Robins, Kaplan, Miller & Ciresi—RKMC for short.

Several Twin Cities personal injury attorneys gathered on a pro bono basis (for free) to represent those injured in the collapse. Twenty Minnesota law firms representing more than ninety victims and families became known as the "consortium." Chris Messerly and Phil Sieff were the lead lawyers and the most vocal and visible lawyers who went to bat for survivors at the Minnesota State Capitol and with the media.

Lisa answered all of my questions. We had a right to take care of ourselves and/or our families. It was better to seek representation now instead of waiting. There was no guarantee how long they'd have enough lawyers who had volunteered their services. Last, many of us had expressed concerns that we didn't want to be greedy, or be seen as greedy, if we weren't hurt as badly as someone else. Did we realize that we were talking about real financial concerns? For example, how much will each survivor be faced with, for medical expenses—especially long term, over the course of a lifetime? Will we be able to recover? This was the tip of the iceberg.

I wrote to the group, one of my almost daily emails, to share what I'd learned and pass on contact information. On a personal note, I wrote, I felt good about recommending them, and I urged people to get all the facts about their specific situation so they could make an informed decision. I wrote, "What do you have to lose?"

49

In Ohio one in four bridges from an inventory of forty-two thousand—the second-largest inventory in the nation—are structurally deficient. As the sun rose over the Grand River (east of Cleveland) on a May 24 morning in 1996, four gusset plates on the Interstate 90 Bridge (built in 1960)—a nonredundant steel deck truss like 35W—bent and "virtually crumpled" on two parallel trusses "nearly simultaneously." Painters had been sandblasting the bridge to prepare the trusses for repainting, and when a truck crossed, they heard a loud bang. The structural failure caused the span to noticeably sag "three inches down and four to the side." The Ohio Department of Transportation inspects bridges annually. The previous November, I-90 had been sound. Bridges are overdesigned to military specifications, to handle excessive loads. Gussets are typically the least likely part to fail. But after years of saltwater running down the diagonal compression members, corrosion had worn the $7/16$-inch gussets dangerously thin. The bridge remained closed for five and a half months. There was no loss of life.

The Federal Highway Administration (FHA) concluded that the cause was a design error in calculating the thickness of the gussets. The incident taught Ohio Department of Transportation officials to inspect gussets more carefully, looking for corrosion and signs of buckling or bending. The lessons rippled through Ohio, the FHA researched the incident, and it was written up in *Civil Engineering* magazine. In August 2007, after the I-35W Bridge collapse, the Federal Highway Administration defended the decision to not sound a national gusset alarm after the I-90 Bridge failure in 1996. As reported in PD *Extra*, an offshoot of Cleveland's daily newspaper, the *Plain Dealer*, "'It was an external force that was introduced that caused the plate to fail,' said FHA spokesman Ian Grossman. 'It was a sequence of very specific actions.'"

The FHA blames the gusset thinning on repainting prep—a typical procedure in bridge maintenance (it inhibits corrosion on steel that accelerates over time, by the presence of moisture, and especially on structures near bodies of water). I'm not an engineer, inspector, politician, or state agency. I'm a survivor who's tasked herself with learning. I'm a survivor who understands enough to realize that the Ohio explanation fails. It fails to take into account that the "external force" was one overloaded yet measly truck, a bridge deck absent the weight of rush hour traffic. And that bridges are overdesigned to military specifications.

50

I was thinking of going to see the bridge to see the wreckage in person, but I didn't make it there. Seeing the collapsed bridge on the news, it's so sterile. The public wasn't allowed to see the wreckage when there were cars there, once all the people who were killed were recovered. The area was blocked off with barricades and police tape. Survivors at group all wondered why. Were they trying to control public reaction? But when we were allowed back, they talked about how strange it was to visit, with people and kids all around. Hearing them talk, one survivor wanted to yell: "I was there. This is what it felt like for me—others had it worse!" But instead, the survivor said, they kept moving because "at the end of the day they would never understand anyway."

8. Bridge wreckage seen through chain link. © 2007 Kimberly J. Brown.

51

Third Fracture Critical Bridge Inspection: July 1996

Same inspectors listed previously. They note again: "Several of the bolts connecting the stringers to the floor beam truss are 'working,' loose, broken or missing. Several of these should be replaced." There are eleven bolts in eight panel points listed, including our "star" missing bolt at stringer #13, span #6, panel point U8, northbound: "One bolt missing, one nut missing (the bearing block has rotated)."

52

At my job, where I worked as a technical writer for the Carlson Companies, I was invited to be the keynote speaker preceding Marilyn Carlson Nelson at the company's United Way kickoff. United Way agencies like the Red Cross, United Way 2-1-1, and Salvation Army were helping people affected by the bridge collapse. I accepted the invitation, hearing myself say yes as if being asked to speak were a common occurrence.

It was an honor, but now I had to write a speech, explain how United Way agencies had influenced my life for the better, and stand before hundreds of colleagues and my boss.

That day a lawyer named Wilbur "Wil" Fluegel—who represented nine of the children from the school bus pro bono—agreed to assess my potential case. Also that month, I agreed to an interview for a TV special called *Beyond the Bridge*. At local station KSTP-5 I walked with other survivors into a brightly lit studio. Collapse images on oversized canvases hung behind rows of chairs.

"Whoa," I said, noticing one of the images in particular.

For the first time I saw myself with Kelly, destruction surrounding us, as we waited to get off the twisted broken center span that lay in the river like an island. Here was further proof, a month and half later, that it wasn't just a bad dream. Often this all felt too implausible. Seeing it left me feeling I must continue trying to make a difference. Why, I couldn't explain. I just had to try.

On October 4, 2007, I met my attorney, Wilbur Fluegel for the first time. Wil would represent me and nine of the children from the school bus pro bono. His office was in downtown Minneapolis, on the thirty-fourth floor of the high-rise towers at 150 South Fifth Street.

The elevator ride terrified me. When I walked into his office, nothing but windows overlooked a canyon of skyscrapers. My mind showed me an image of an impressive wind, a tornado, a vortex that smelled of heat and combustion, sweeping in and ripping everything away. The floor, his desk, the huge stacks of files, scads of legal books, him, me—ripped from that space in a violent upheaval into a void of blue sky. Up so high, no one heard our screams.

Blink hard—return to now. For a time people cited statistics, like they'd all received the same script, trying or thinking they were helping. "Statistics show that the chances of a bridge collapsing are . . . ," and I would stop them. "Don't tell me about statistics." My social filter stopped me from blurting, "Why don't you shove your statistics?"

We're not a developing country. We're supposed to have standards: engineering rules and cross-checks and triple checks and minimum requirements for safety. The bridge collapse rendered statistics useless. It only takes one bridge—one place, one perfectly timed moment of horrendous serendipity, God calling you home, dumb luck, omens or hexes, your time regretfully run out—and you can become one in a million. A statistic.

As time passed, I grew too tired to correct people. At the "statistics show" speech I began responding with, "Uh-huh." The introverted side of my personality became dominant. What I was going through was entirely invisible, and there were few people who understood what I was thinking about. Catastrophe teaches us to notice. Whether it's catastrophe on a personal level (think cancer) or the communal catastrophe (think Katrina, 9/11), we control very little in our lives. In the immediate aftermath, Conscious Me was superstitious and paranoid. Conscious Me thought being fit and healthy is a curse. Being healthy makes you want things. Makes you hopeful, makes you possess a positive mental outlook, makes you unprepared to be struck down. When you're blindsided, there's nowhere to go but down.

53

When a bolt "works," this means it is vibrating and moving due to stress. Stress is the load on a structure, element, or component. It is usually described in "pounds per square inch," or psi. Think of something one inch square with 1,000 pounds resting on it. The element is under "1,000 psi compressive load or stress." If the part is pulled with 1,000 pounds, it is said that it is under "1,000 psi tension" stress. Of course, the part must have equal and opposite forces in order to remain in one place. If the equal and opposite forces are not precisely opposite each other, the element is said to be in "shear." Shear stress is expressed the same way as compression and tension: "1,000 psi shear."

54

Sometimes we must accept that we cannot solve another's woes. Sometimes all we can offer is to listen and bear witness. Other times we must step in and act. We must use our talents for the benefit of the greater good, whether that's for the good of two people or two million, when something larger calls.

On October 8 I stood before an audience in the marble rotunda at my place of work, at the Carlson Towers, preceding Marilyn Carlson Nelson, and gave a speech to kick off Carlson's United Way campaign. Leading up to the big day, Rachel listened, gave pointers, and encouraged me to practice. The morning of the speech, trying to rehearse, my tongue stuck. I left for work with a sense of doom. I was going to bomb big time. A few hours before the momentous debut, I locked myself in a conference room and delivered the speech to a panel of windows overlooking a fountain, a lake, and two major freeways that weaved a continuous traffic flow. It wasn't until I remembered the thirteen that I felt: I can do this.

Rachel took off work to see me speak. In my speech I connected experiences with the United Way growing up—that I hadn't realized were part of my life—to the agencies that helped people after the bridge collapsed: the Red Cross, United Way 2-1-1, and Salvation Army. I talked about "community," how I wanted to live in a vibrant place where neighbors care for each other. Afterward I couldn't believe I'd done it. I gave a speech in front of hundreds of people. Amazing how fate works, how I can be so much more than I thought. I received an email from a colleague that said, "I was very moved by your speech, and as a result, I made a pledge." Emails like that continued to trickle in. It was a great feeling. Jennifer Matheny, my manager at Carlson, said, "Always turn a negative into a positive." We don't realize how much we have until it all almost gets taken away. That's the paradox. Others expressed interest in having me speak again, an interesting twist since I worked hard to become a writer, not a speaker.

I had attended the support group each week since August 24. It had grown steadily. When I told people about the group, they asked, "Does it help"? I never knew how to answer that. I had mixed feelings. To know others who were there, that instant recognition, not having to "explain," I was grateful for these connections. But I felt anger for the pain and hardships people had to overcome and faced. Anyone's story of injury, death, or loss could have been mine. I mulled it over, imagined what each person went through. So, did it help? I had to go to group. It was hard but also affirming and, occasionally, a gift to not feel alone.

55

Fourth Fracture Critical Bridge Inspection: August 1997

The reports say the bolts connecting the stringers to the floor beam truss are "working" loose and breaking. There were at least twenty bolts that needed replacing. The inspection report says it will be done as part of the upcoming painting/repair contract, but the repairs weren't done. The bridge was cared for with deferred maintenance.

Part of me wants to say, See, the bridge collapse was caused by maintenance issues. But it's more than that. The broken bolts alone are not what caused the bridge to collapse. It's not black-and-white. Like many things in life, it's a combination of factors. The final factors I won't know for nearly a decade. If the bolts transferring the bridge stresses to its members were repaired and the bearing blocks seated properly, working in concert to let the bridge function as it was intended—if only . . .

A useful analogy might be this. If you have a heart attack and you die because you have a heart attack, while some may say the cause of death is heart attack but for years, perhaps most of your life, you ate poorly, drank alcohol in excess, never exercised, and did all of this knowing you had heart disease in your family history, then all those years of abuse and neglect contributed to your demise. Basically, the bridge was in bad shape with undue stress that could've been ameliorated with known fixes. If it had been in good shape, maybe it could've stood, potentially, maybe had more resilience. Based on the bridge's lifetime evidence of disrepair, it all contributes.

56

The following morning after the speech, I drove to work and couldn't stop thinking about the previous evening's group. All we heard about on the news was the money it would take to rebuild the bridge. The 35W Bridge would be replaced quickly for $225 million plus a hunk of bonus money, $25 million, to complete the work fast. At this time it wasn't known that the cost to rebuild would be closer to $400 million—a new price tag that included "the cost of rebuilding the bridge, potential incentives for the contractor to finish early, and the cleanup of the collapsed bridge." According to a September 15, 2008, MinnPost.com article, when all was said and done, "construction costs totaled $234 million, not including the bonuses Flatiron-Manson is expected to receive for completing the project ahead of schedule. Gov. Tim Pawlenty said the closure and detours cost the state's economy an estimated $400,000 a day in added travel and other expenses." But there had been no mention of how to help the people. I pulled out a notepad, balanced it against the steering wheel on an endless stretch of straight freeway, and jotted chicken scratch notes.

Those notes turned into a letter, which I sent that afternoon to every member of the Minnesota House of Representatives and Senate and our U.S. House and Senate representatives.

Then I contacted a reporter from the *Star Tribune*, Pam Louwagie, whom I had interviewed with previously in a story called "Aftershocks," to see if they might run a story about my letter and thereby bring attention to the situation. Pam had taken great care with survivors. She had called me and other survivors featured before "Aftershocks" ran to give me an opportunity to hear the story, which impressed me. I told Pam about the group. It had been growing, and several of the more injured people were really struggling, which had brought out more issues. Director Margaret McAbee and chaplain Frank Thell, the support group facilitators, had been encouraging survivors to write letters to our representatives. No one had acted yet, which was frustrating. But we were all new to this whole politics thing. The lack of action was probably due to unfamiliarity (how do we do this?), and people were overwhelmed, injured, grieving, and hurting—they'd almost died from a bridge collapse. Once people wrote their letters, the support group facilitators encouraged us to share them with family and friends and collect them in a packet that could be viewed with the strength of the group.

I told Pam that I believed I was the first to write and send a letter. I hadn't shared my letter with the others yet, but I would do so as soon as I got my letter sent to each legislator. I expressed my hope that my letter could be printed as is or in part. Excerpts from "a group letter" could be good as well, I said, if the others could react quickly. Pam's article, "I-35W Bridge Survivor

Vents Outrage; A Survivor's Letter," was published in the *Minneapolis/St. Paul Star Tribune* newspaper online on October 11 and in print the next day.

Two months and eleven days after the 35W Bridge fell, why have no legislators stood up and said, "Let's take care of the survivors"?

Why does the State of Minnesota have 27 million dollars to reward a contractor to build a bridge with an emphasis on speed, but zero dollars for the survivors? Instead, survivors must navigate the bureaucracy of a nonprofit system that is run on the grace of Minnesotans' generosity.

Why do Minnesota politicians pat themselves on the backs, saying that the disaster wasn't as bad as it could have been, when this bridge should never have fallen in the first place?

Why does Minnesota put hundreds of millions of dollars into sports stadiums, but months after the disaster, the survivors are left to fend for themselves? Why have the state or federal government—who was responsible for the bridge—given $0 dollars to help victims recover?

Instead, I listen each week to people's pain, as they worry about how to make it from day to day. I've been going to the Bridge Collapse Survivor's support group each Wednesday since August 24th.

Why am I hearing, "My spouse can't work, we have no money coming in." "We don't know how we are going to pay our mortgage." "I got a medical bill for forty grand; that's the tip of the iceberg." "I lost my job." "I can't drive myself to doctor appointments because I'm hurt." "We need to hire a housekeeper because I can't do the things I used to." "My spouse has become my caretaker." "I can't work." "My spouse can't work." "We have no money coming in." "I can't sleep." "I jump at the smallest noises." "I'm always sad." "Our money is gone." "We've been forgotten."

I want to hear, "My buddy died, but Minnesota is taking care of me." "My medical expenses are covered, I don't have to worry about them." "My job is protected."

"My mortgage is paid." "A caretaker is provided." "Transportation is provided." "I have help with day to day living." "I can see any mental health professional."

I want to hear, "I paid my taxes." "It was everyone's bridge." "It doesn't come out of my pocket." "I can just focus on getting better."

Before you build that Memorial below the bridge, where none of us want to go, how about someone take care of us first? How about someone ask us what we want?

United Way agencies—such as the Red Cross, United Way 2-1-1, and the Salvation Army, which are run by donations and volunteers—should not be the only one to bear the responsibility of helping those who survived.

Survivors have been frustrated and angry with the red tape and lack of response, and who can blame them? Yet, I maintain that our anger is misplaced—where is the State of Minnesota and Federal government's culpability? Why is a nonprofit organization the only agency helping victims? This isn't even their bridge. If Minnesota won't bear some culpability for this failure, how about some of that $27 million that Minnesota was willing to give away be established as a fund for survivor recovery, and thereby say, "We just want to take care of you."

I want these innocent people to have all of their medical, all of their insurance, all of their mortgages, all of their lost wages, all of their counseling taken care of by the entities that were in charge, or partially in charge, of that bridge. And not just today, but months and years from now.

It's time to slow down and back up. Your new bridge is going up too fast. You still have a huge mess from the old one. Fix this.

I forwarded a copy of my letter to the group. I told them that I'd sent it to every member of the Minnesota Legislature. I included links to the legislative district finder, a website that matches where a person resides with their government representatives. I said: "I am really sad that everyone is going through such awful stuff. Now I hope EACH OF YOU will write your letters."

57

Our star missing bolt remains. Panel point #8, east truss, pier #6, stringer joint: "One bolt missing, one nut missing (the bearing block has rotated)."

Two inches in thickness were added to the deck with the addition of another layer of pavement, a 15 percent weight increase. The bolt was under stress from tightening when the part was assembled. The cars and trucks as well as the stress from heating and cooling from 100 degrees to minus 40 degrees caused loads to cycle millions of times, and the bolt finally failed due to fatigue. The cycle of these loads also caused the block to "walk," or work, out of place.

58

After my letter appeared in the *Star Tribune*, good things started happening. I was invited to appear on Twin Cities Public Television for the public affairs show *Almanac* to talk about the bridge survivors' needs. I interviewed with hosts Eric Eskola and Cathy Wurzer in a setting much like a dinner table. The show was live, which made me extra nervous. If I messed up or got tongue-tied, no editing. Everyone would see. Before the interview my skin itched. I felt hot. At one point during the interview, Cathy Wurzer asked, "What's it like to be an advocate?" I was pleasantly taken aback. No one had referred to me with that word before. I stammered something like, "If my words could help someone, that would be a wonderful outcome."

Senator Amy Klobuchar's assistant called me the same evening that my letter and corresponding article were published in the *Star Tribune*. I paced freshly mowed grass at the soccer field where my oldest niece practiced, as I explained the situations of bridge survivors. Senator Norm Coleman wrote a letter that was sympathetic but directed me to social service programs—the Low-Income Energy Assistance Program (LEAP)—or "available resources," which I didn't understand. What did an energy program have to do with recovery from a bridge collapse? Basically it said, Sorry this happened, but you can sign up for reduced bills this winter. Coleman's response so underwhelmed me that it was hard not to feel angry. But this irritation drove me. There was much work to do. Other responses, from both sides of the political aisle, were compassionate and sincere in their efforts to find solutions.

After the interview my hives erupted. Amy Mattila—my massage therapist and a wise, intuitive friend—said I could try a bath of Epsom salt and eat cool foods. She said the body's psychosomatic swelling is its attempt to release cortisol, an adrenaline hormone in response to stress—especially during the body's flight-or-fight response. The morning after *Almanac*, I sat up slowly, waited for my back to stop throbbing. Reflected.

Later, at work, I got a call from a man named Joshua Freed from the Associated Press to ask if I would be interviewed. Leaving my desk, I scrambled to find a private place to talk. I ducked into the IT department's electronics storage closet. Surrounded by shelves of defunct computers, mice, keyboards, and cords, I shared the information with a national, even worldwide, media outlet.

59

Fifth Fracture Critical Bridge Inspection: September 1998

At stringer #13, span #6, panel point U8, northbound: "One bolt missing, one nut missing (the bearing block has rotated)."

60

On October 17, as a result of the *Star Tribune* article, I received an email from a relative of one of the people killed in the collapse. He contacted me via Pam Louwagie, the reporter from the *Star Tribune*.

The experience of reading his email felt so intense, I had to stop a few times to get through it.

He told me that he appreciates that we were bringing attention to the lack of assistance. *I can't believe he found me*, I said to myself. His name is Bob, and he writes in all capital letters.

MY NAME IS BOB ROSS AND I GOT YOUR E MAIL ADDRESS FROM PAM AT THE STAR TRIBUNE. I MET HER SATURDAY AS SHE CAME TO A SERVICE FOR THE VICTIMS THAT DIDN'T SURVIVE THE 35W BRIDGE. I HAD E MAILED HER FRIDAY AFTER I READ HER ARTICLE ON THE FRONT PAGE OF THE STAR TRIBUNE AND AFTER I HAD READ YOUR LETTER YOU SENT TO THE SENATORS.

MY DAUGHTER'S HUSBAND, SCOTT, WAS KILLED ON THE BRIDGE. HE WAS COMING HOME FROM HIS JOB AT CAPELLA UNIV AND THEY WERE GOING TO PICK UP THEIR NEW CAR. SCOTT WAS RUNNING LATE AT WORK AND CALLED BETSY ON HIS CELL PHONE AND SAID HE WAS JUST TURNING ONTO THE BRIDGE AND TRAFFIC WAS BAD. THEY WERE MARRIED FOR ONLY TEN MONTHS. THEY WERE MARRIED AT THE CARLSON CENTER LAST YEAR ON SEPT 30. IT WAS A PERFECT DAY AND A FAIRY TALE WEDDING. I HAD EVEN HAD A HORSE AND CARRIAGE FOR MY PRINCESS AND HER PRINCE. IT WAS A TOTAL SURPRISE TO THEM AND EVERYONE ELSE. WE HAD WORKED WITH DAWN AT THE CARLSON AND SHE DID A GREAT JOB WITH HER STAFF. AFTER SCOTT WENT MISSING SHE WOULD BRING FOOD BY EVERY WEEK WITH HER FAMILY. IT WAS TOUGH WAITING NINETEEN DAYS FOR THE NEWS WE KNEW WAS COMING BUT DIDN'T WANT TO HEAR. I COULD GO ON BUT I JUST WANTED TO THANK YOU FOR WRITING THE SENATORS.

IT IS A SHAME THE LACK OF ANYTHING FOR ANYBODY. YOU INSPIRED ME TO START WRITING. I KNOW MY DAUGHTER DOES NOT WANT TO BE IN THE PRESS BUT IT IS TERRIBLE HOW ALL THESE PEOPLE THAT HAVE PROMISED SO MUCH HAVE DONE NOTHING. I DON'T KNOW WHAT YOUR EXPERIENCE IS BUT THE RED CROSS IS THE SCAPEGOAT. EVERY AGENCY THAT HAS BEEN COLLECTING MONEY REFERS YOU TO GO TO THE RED CROSS. THEY DO NOTHING EXCEPT MAKE YOU FEEL BAD. THEY GAVE BETSY $2,000 FOR FUNERAL EXPENSES BUT THE OBITUARY ALONE WAS $1,250.

WE ARE NOT LOOKING FOR HANDOUTS, BUT SOMETHING WOULD BE NICE TO COVER THE EXPENSES THAT YOU TALK ABOUT. TODAY IN THE EDITORIALS A LADY WROTE ABOUT THE ARTICLE AND IT IS TIME TO KEEP THESE ACCOUNTABLE.

THANKS AGAIN,

BOB

I sat at my desk at work and read the email. To keep from screaming, I slapped my hand over my mouth. Tears drizzled down my cheeks. Had

I been at home and not in public, where I had to keep myself together, I would've bawled. This was so much validation, so much proof, that I could help. Something good finally happened.

I wrote to the group, "All of you—know that you don't have to feel invisible anymore. And let's all keep remembering that there are others out there that we haven't met yet."

61

On October 18 WCCO-TV tonight at 6:00 p.m.—they did a whirlwind interview on a speedy turnaround. I think they interviewed Lindsay and me; not sure if they were able to get ahold of anyone else. Also on this day Bob and I corresponded again. He forwarded me a letter he wrote to Senator Dick Day, and he also said:

> I WANT YOU TO KNOW THAT YOU ARE A REAL COMFORT TO HEAR FROM AND YOU ARE LIKE ERIN BROCKOVICH (IF YOU SAW THE MOVIE) IT IS ONE OF MY FAVORITES.
>
> ONE THING THAT I THINK MIGHT BE A BIG HELP TO PEOPLE IS IF YOU HAD A LIST OF KEY OFFICIALS OR A WEBSITE TO GET THEM FROM, IT WOULD MAKE EMAILING OR SENDING TO THESE PEOPLE EASIER. MY KNOWLEDGE OF GOVERNMENT IS ZERO. I DON'T KNOW A HOUSE FROM A SENATE.

I wrote, "That is one of my favorite movies," and, "Try to do something good for yourself, even if it's just a little thing."

I didn't know what to say, as a survivor, to this person who had endured the worst, most irrevocable forever loss imaginable, to lose a loved one.

"Go get a hot fudge sundae or whatever makes you happy. You have to take care of yourself and stay strong for Betsy."

> IF IT WASN'T FOR YOUR GOOD WORK, [OTHER FAMILIES] WOULD NOT HAVE CALLED OR SHARED THE EMAILS WITH ME. SO PEOPLE ARE STARTING TO TALK AND ARE TIRED OF HOLDING BACK. . . . THANKS FOR YOUR ADVICE. I WILL HAVE A HOT FUDGE SUNDAE

AND WILL WATCH THE NEWS. THANKS FOR ALL YOU DO. IT HAS BEEN VERY HELPFUL.

62

Sixth Fracture Critical Bridge Inspection: April 1999

Panel point #8: [1994] One bolt is missing, and the nut is missing from the other bolt—the bearing block has rotated. [1999] Missing bolt replaced (bearing block still rotated)—photo.

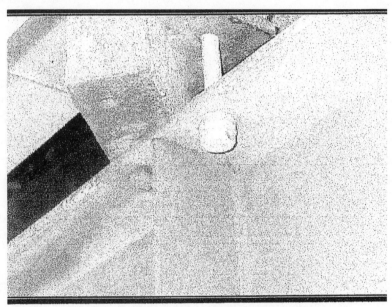

Panel Point #8 East Truss Stringer #2

9. Broken bolt and bearing block rotated. Bridge 9340, Fracture Critical Bridge Inspection Report, April 1999. Minnesota Department of Transportation.

63

At a doctor's checkup the nurse announced, "Five feet one and a half." "What?" I said, ducking from the ruler, thinking she was joking. "I'm five feet two and a quarter." I returned to be measured. She confirmed, this time holding the marker in place. I have lost three-quarters of an inch in height. I used to be five feet two and a quarter. Could I regain this height? I looked into therapies, including an inversion table. I told friends that I hang upside down like a bat. I also chose to do other things that wouldn't make the physical injuries worse. A return to running felt terrible. Sharp stabbing pain in my knee. So I stopped driving a five-speed. I don't climb stairs. (Confession: I'm not really broken up about the stairs.) My chiropractor teaches me physical therapy I can do at home. All of it is work, but I can do these things.

64

After I sent my letter, survivors submitted theirs. A few survivors, whose emails had previously fallen on deaf ears, began receiving responses from their senators. I sent frequent emails to those in the support group and said, "The NATION is now hearing of the difficulties survivors have been dealing with. As your stories continue to be told, greater understanding will be had by all. So keep your letters going. It's making a difference." This was the grassroots mentality that propelled us.

A few weeks later, on October 25, survivors and family members of those killed participated in a press conference to discuss the need for a bridge bill—a fund—to benefit those harmed by the bridge collapse. I testified, along with six others, before a small panel of about eight legislators from the Claims Committee, and media—all of the local channels, Senate TV, and others—pointed their cameras to document the proceedings.

The press conference took place in the Heffelfinger Room at the Red Cross Building, a short distance upstream from the collapse site. Survivors and families testified about their $20,000 no-fault personal injury protection (PIP) automobile insurance policies not getting you far in the hospital when something unprecedented happens (like a bridge collapsing)—survivors in the hospital had already used it—and about worker's compensation replacing two-thirds of income, which for one survivor was less than $300 per week. The Minnesota Helps Bridge Disaster Fund, established to collect charitable contributions from the community, had gathered over $600,000—yet one survivor alone had medical bills in that amount. One by one women and men came to the microphone, some in wheelchairs and back casts, and spoke of their losses, whether physical, mental, emotional, economic, bereavement, or all of the above.

One man took the impact of the collapse with his face. His compact red car fell nose first: jaw broken in three places, face and nose bones shattered, eye tissues bruised, tubes in brain and lung, broken leg, broken arm, colon and abdominal wall severed. Another woman's legs had been shattered below the knee. Six surgeries so far to repair them, more to come. Another survivor's car sunk to the bottom of the river with her in it. She still doesn't know how she got out but remembers being trapped. Others came to the microphone and described when their mates didn't come home and they knew something was terribly wrong. In front of all of these people, in front of the media, one by one we described with rows of camera lenses pointed at us, to the click of photographers taking pictures, the terror of falling with the bridge, of MRIs, post-traumatic stress disorder, the raw weight of grief for loved ones who would never cross that bridge or come home to hold their children again; and survivor benefits held by families, neighbors, and concerned friends that were of course a blessing but couldn't be sustained long term. Everybody in the room was crying.

65

After the hearing at the Red Cross Building, my attorney, Wil, said the day was a great success, but it was only the beginning. Additional hearings would be scheduled in the coming months until February, when the legislature convenes.

To listen to the stories of when all was lost or could've been was overwhelming. I came home and couldn't sleep, couldn't socialize or enjoy life. Every time I started, I thought of them and felt uncomfortable in my skin.

Still getting hives, I would awaken late, rising lethargically from Benadryl. My neck out. Sharp pulses of electrical-like pain shot up the back of my head. My lower back throbbed. I felt old. In the mirror I had new wrinkles on my forehead and gray hair by my temples.

"Do you feel like it's over, that you're safe now?" Kris, the new therapist I started seeing, asked. "Not really," I said, describing how I'm frequently nervous, thinking that something else is coming. A few days after the collapse, someone at work had said, "Bad things happen in threes." I was appalled but kept it to myself. *What a thing to say to me. Like the bridge collapse by itself wasn't three bad things packaged in one?* Others have said, "You've got something big planned for your life." How do they know? Do they know that's a lot of pressure? Part of me felt bitter. Maybe I was supposed to be ordinary and just live my life like everyone else.

Kris said, "Injuries from being 'run over by a truck,' people get that. Emotional, psychological injuries—people don't believe what they can't see with their eyes." I shot her an exasperated look. She suggested two quick strategies for when I'm scared. Breathe in and out twenty-five times. This shifts your respiration, calms you down. I looked at her suspiciously and wondered if my cynicism showed. Then she said, "Try writing out your experience."

In describing PTSD, Kris said that any minor thing can upset one's equilibrium. "Things you'd normally take in stride, right now you can't deal with." Yep. I would get easily overwhelmed and depressed. I said, "Don't make me think. Don't make me organize. Don't make me solve anything." The most normal places felt scary. And nobody knows because when I try to explain, they look at me with confusion or pity, and who needs that? Kris tells me to think about what I've lived through. It's risky to think too far ahead. "You were going to a soccer game," she said. The more she hears me talk, the more she thinks I really am depressed in addition to the post-traumatic stress. "Even the anger, the irritability, can all be a sign of depression. And the sleeping, just sort of slogging through . . . nothing really feeling good." She told me she wants me to "try to have more fun. Try to do things you enjoy."

I had no idea how to do what Kris advised. Survivor's guilt dominated my life, although I never called it that until one day in therapy—describing the stress, my grief, my exhaustion—I blurted out that I wanted to know that I "deserved to live." It stopped the session cold. I gasped. My therapist oohed.

"Where did I get this?" I asked, stunned. This is where past trauma comes in.

I thought I was going in for counseling for this event. When this therapist asked me all about my childhood, my response was, "Why?" I was here for a specific discrete event—*this* event. I didn't need to dredge up the past, and not because I just didn't want to talk about it but because I'd invested in therapy in the past and was better for it. Yet over the months ahead, I'd discover that, more than I wanted to acknowledge, the past dictated the future; understanding it was a kind of key.

Once, at a massage session, Amy told me, "Wow, you have some major survivor's guilt going on."

I said, "No I don't. Everyone else has that."

"Yes, you do. You can stop. You can do your work, do your bridge stuff. Then come home for dinner and stop for the night. It's okay to work, to care, to worry, but at some point you have to turn it off and just be with Rachel. Be at home with your family and just be."

This struck me as a foreign idea.

"But I can't stop. I think about it all the time."

How could I interrupt this momentum? Guilt allowed me to act in extraordinary ways. To be outgoing when normally I was shy. To speak when normally I'd let someone else do it. To write and spend my free time researching contacts and editing and perfecting each word. To visit survivors in the hospital. To be socially outgoing with legislators and lawyers, probing them for answers and information, presenting ideas and asking to be involved.

Interviews were extensions of public speaking. I set aside nerves and reclusive tendencies. Pulse quickened. Studio lights hot. Meeting reporters for interviews, temperature, wind, sun, the elements, the camera, even the proximity of the reporter or the comfort of a chair I sat in, played a role. Whatever the situation, it was a test. I shook, my palms sweat. Clothes could be a barrier. How would I look? Did I sound all right? Would I know what to say? Insecurities stewed within like a simmering soup. On the set of another public television show, I sat across from an interviewer. Cameras panned the room—the stage elevated on a platform, two chairs, a table in the middle, a plant, the interviewer and me.

"What was the collapse like?" the interviewer asked.

Thrown—I'd come to talk about the condition of Minnesota's bridges, not the moment I thought I was going to die—I turned into a mousy, dreary person. I looked and sounded nervous, my voice a whisper. I should've stopped him, taken control, and said, "That isn't why I'm here." Reviewing a tape, it wasn't until the more academic questions started about the status of bridges that I recovered an air of confidence. At another interview—this one by the Mississippi River on a 95-degree day—I stood with a reporter and a videographer. I looked at the camera lens and felt beads of sweat on my upper lip and forehead. Under full summer sun, did the camera catch each droplet glistening in the foreground of a new bridge being built?

In the coming months important decision makers took notice. I agreed to multiple interviews and discussed the possibility of a victims' fund (like after 9/11), sponsored by House representative Ryan Winkler. Some legislators said we had to wait until blame was determined, but I maintained that many couldn't wait. Fire warmed my belly like I'd knocked something vital into motion. These people's lives might be changed for the better because I spoke up. The continuing momentum grew addictive and poetic, the metaphor of my childhood versus the present. As a kid, my natural inclination was to be silent in the face of life's challenges, willed by what my parents thought was best, and standing behind my mom, who was usually the one who spoke.

My mom lived with type 1 diabetes: chronic, incurable, life threatening. She was diagnosed when she was twenty-six and lived to age sixty-four (she passed in 2004), but the losses she endured were catastrophic for her and her spirit. Growing up, she was my best friend, and I wished so badly that I could've fixed things for her. Because of my dad's fame and the media attention that sometimes went with that, our family system was to not talk outside of the family about loss and difficulties. I remember vividly Mom looking in the bathroom mirror, pulling a curling iron through her bangs, telling the young me, "This stays in our family." And it was then that my unofficial media education began. Those were years when our family would participate in public events like NFL charitable auctions, diabetes fund-raisers, and media interviews, and we learned firsthand the benefits and dangers of being in the public eye.

What I didn't know then was the level of guilt and helplessness that drove me to act. Could I let myself celebrate my survival? With time I realized my actions were all ways to be a hero for innocent people, like my mom who had been hurt by a cruel disease. This need to speak up came from somewhere. Powerless as a child, I had been silent. But now, as a collapse survivor, I tried my hardest to turn this trauma into a new story. By sharing with others, I empowered myself. Everything I was doing was in direct opposition to my inclinations, a complete growth trajectory away from the patterns of my childhood. As an adult, I was speaking—and I had something powerful and important to say.

66

In 1998–99 another 5 percent was added to the weight. Fatigue is failure that happens when the stress varies, or "cycles," partially or completely. Fatigue failure is like that (but not exactly) paper clip mentioned earlier, the one that's bent until it breaks.

67

I dreamed Rachel and I moved to an apartment in New York. I saw a skylight in the closet and asked Rachel, "How tall is this building?" She shrugged. Apparently, our new place was on the fifth floor. I left the apartment, and I thought what I needed was one floor above or below, so I got in the elevator, where three other women were riding. I saw a few buttons but no 5. I said, "No," signaling I wanted to get out. But before I could, a female passenger pressed a button, and numbers began changing rapid fire. The digital display held four digits, which is when I found out that the building had 1,500 floors. The compartment shook violently, my stomach was gone, and I grabbed my head and screamed, "*No No No No!*" I thought we'd punch through the top when the elevator finally slowed, and we were weightless. People exited, the doors slammed shut, and then we had to go back down, over one thousand floors.

68

I rushed to find a meter and circled the blocks several times before I found one. When I called Wil to let him know I might be late, his paralegal Nancy answered. Previous meetings, at my request, were held on the second floor.

I said, "Oh, by the way, I'm going to brave the elevators."

"Good for you," Nancy said, "You're getting better."

I pushed myself, reasoning that if I expected to go on a decent vacation again, ride on a boat, or go anywhere, I had to start getting used to moving, to machines that transport us places. As I walked to the skyscraper, turning the corner toward its smooth red steel and brick, I started shaking. Crap. I was going to be extra brave, but PTSD isn't just in the head; it's in the body. Inside I crossed the courtyard to the double banks of elevators and stopped at an information desk.

"Where's the restroom?" I asked the clerk.

"Do you have an appointment?" she asked.

"Yes."

"Well, they'll have them on their floor."

Silently, I objected. But the elevator will make my stomach drop out and... and... there was more involved with the Sisyphean task ahead, but I didn't bother to explain. How would I? I smiled halfheartedly and said thank you.

Here I go. I inhaled and pushed the triangle UP button. I waited. Ding! The car arrived. Double doors cracked open, and I stepped into the elevator car onto the low-pile carpet. I scanned what seemed like hundreds of buttons, found 34, and pressed it. The car moved as expected, pulling up hard, and my weight pushed through my toes. The compartment shook like a cocktail shaker. On the way up I felt lighter and tensed my stomach. To thwart terror, I did Amy's coping trick: observe a surface with touch and texture—a temporary way to root the senses in the present instead of an imagined future. I concentrated on the bronze railing. Quick, I observed it! I rattled off words nonstop like a rapper. *This is smooth this is cold this is shiny this is metallic this is pretty.* Then the elevator slowed, and I felt weightless again. The drone dulled. Ding! Floor 34. This is good. The car rose as it was supposed to. No surprises.

I hurried out and headed to the restroom. My mouth dry, I felt like I hadn't eaten. Low blood sugar? No . . . my body hummed. Amazed, I realized I knew my body so much better now. This was adrenaline. Like on that day. Now I could identify this feeling with words. I composed myself with a mental pep talk. When Wil led me into his office, I noticed all the shades on the wall of windows had been drawn. Surprised, I pointed and asked, "Did you do this for me?" At the same time, he said, "Look, I drew the shades for you." I was touched and told him so.

With folded hands, a white-gray beard, and perfect legal sentences, Wil explained lingo about the settlement with the State of Minnesota—options, risks, and benefits. He spoke of the Minnesota Insurance Guarantee Association, or MIGA, telling me about a client whom he represented who had been injured and was now a paraplegic. A sense of claustrophobia crept in, and I wondered why I hadn't brought water. I coughed feebly, sounding like a cat with a hairball.

I asked, "Can I take a break, or do you have some water?"

Wil practically leaped from his chair. "I'll get you some," he said, disappearing around the corner.

Try to relax, I told myself, but my throat felt like a desert. Wil returned with a glass of water, and I thanked him and gulped it down. I thought about the thirteen surviving families. What must it be like to take this in? Dealing with a settlement for one, where once there were two.

69

The elevator at work malfunctioned *with me in it*. I was on the twelfth floor and pressed 5. The button didn't light. I pressed it again. Still nothing. By now the doors had closed. I pressed 5 insistently. Nothing! Then the elevator began a slow-motion fall. The board of numbers remained unlit. The car stopped at 9. I pressed the OPEN DOOR button and—nothing. Flailing palms, I was afraid to press something wrong but also afraid to do nothing. The elevator resumed its descent. I fought panic. It stopped at 8. I punched 7, where it finally opened. I rushed out, shaking, heart pounding. I took the stairs two flights to my floor.

I told a coworker, Melanie, what had happened. Shocked at first, she teased, "You'll need to be escorted from now on."

When I reported it, Security asked, "Which one?" I couldn't remember, so I walked back to check. The area had six bays, and on the trim around each door, each elevator had two labels. One for the floor and on the other side a combination of letters and numbers. I called Security back and told them, "That elevator is B3." I spread the word. "Avoid elevator B3."

That evening, at home after dinner, while brushing my teeth, I was thinking about the elevator, replaying it in my mind, when I yelled, "Holy shit!" The National Transportation Safety Board had spray-painted identification letters and numbers on everyone's car. The car I was in was B3.

Rachel said, "Maybe it's just a weird coincidence."

"Spooky is more like it."

Was the universe trying to tell me something? B3? How could it have the same label? How? It was just two days before Mom's birthday. In paranormal stories they say the ghosts aren't 100 percent accurate. Could it have been Mom?

I told Amy about it. She said, "I don't think that you failed or the big God is picking on you. I think that when we start to stretch our wings and learn to fly, it's not always smooth sailing. Falling is part of it. If you're not falling down, you're not learning. You're doing great. You're stronger and bigger than B3. You're simply learning to fly. We should make up something funny for B3 to mean—I think it means bullshit three times."

Another morning, walking the usual way from the surface lot through the loading dock and the underbelly of the towers, the UP button was lit, but no one was there. Strange. I approached the bay to wait, but before I stopped, the bell dinged and the doors opened. The bell echoed as I stepped over the threshold into the darkened elevator car. I checked the trim: B3.

70

Seventh Fracture Critical Bridge Inspection: April 2000

At panel point #8, stringer #2 has loose bolts, and the bearing block has rotated.

Fatigue is the cycling compression, tension, or shear load that may or may not cause deforming, but it can cause sudden failure. Steels are graded as to the stress they can take before deforming or failing under stress. Remember that stress is in psi, pounds per square inch. If the steel is half rusted away, our example of 1,000 pounds is spread over half the area, causing twice the stress (2,000 psi). The L-11 gusset plate was 20 percent rusted away.

71

At support group the facilitators encouraged us to meet with our district representatives and senators. In person. They had to know who we were. They had to meet us, talk to us. Be able to associate a face with a name. When they pulled the dossier for "Bridge collapse survivor Kimberly J. Brown," they'd be able to conjure details—female, midthirties, black hair, brown eyes—recall my voice and even how I dress. But why? None of us understood at first. This felt like yet another hurdle, a bar placed before us, with our injuries—some visible, some not—and we're supposed to jump! I heard this every week and fought a sense of doom. How on earth would we get hundreds of survivors to identify who represents them by district; write to them; schedule a time to meet them; meet them; organize their thoughts; and not chicken out?

In the future the complicated counterarguments, the need to convince, would become more clear. Without going into a lengthy explanation, here were some of the questions coming from other districts. Can we clarify who is really responsible for caring for the survivors? Do we need to determine fault or blame? Do we need to know the cause of the bridge collapse before we can go forward? What structure would allow the state to compensate victims without proof of liability?

But on November 15 I met with my district senator, Patrícia Torres Ray, at Nokomis Beach Coffee, a quaint neighborhood hangout. When customers entered, a cowbell on the door clanged. I found a spot to sit, then changed tables a few times. When I finally settled in, I reviewed notes, kept close at hand. I'd prepared discussion points and questions with specifics. A survivor's bill from Hennepin County Medical Center for $525,000, just for the hospital stay—not for the surgeries and all the "items." Or the significant others who had to stop working because their spouses or partners were hurt and needed someone to care for them. No insurance for it. Or the common situation where no-fault auto insurance had run out. Minnesota law required the limit of $20,000, but those who had been hospitalized had already used it. Within a folder I clutched facts to my heart as I rehearsed which ones to focus on. People would need long-term help, not just short-term. This bridge should still be standing.

I rearranged my personal effects OCD style, handled my bag, coat, pen, papers, keys, cell phone. Then I heard the bell clang as Patrícia—pronounced with the accent on the first *i*—arrived. Originally from Columbia, I guessed her height at around five feet five. Petite, female, and an ethnic minority like me (I'm Korean). Cool. You can do this, I told myself. After introductions we got down to business. As soon as she started talking, I realized she was sharp like my favorite Furi Santoku kitchen knife. Patrícia told me what survivors were up against. The governor and administration were opposed to any dialogue for funding projects. Everything required a source of revenue. There was a concern about setting precedent. Survivors must demonstrate that this was a unique extraordinary event. It seemed like a no-brainer to me, but lawyers who have dealt with all manner of legal cases brought against clients wanted to exercise caution. At that time, with the Minnesota Helps Bridge Disaster Fund, the community had raised roughly $600,000. "What you could do is say, Now we're asking the state to be part of the solution, perhaps match this."

How? I wondered, furiously scribbling notes. At one point I asked her to wait so I could catch up. Ground coffee beans wafted. Whipped cream, in a frenzied dollop in my cup, melted to a thin sheen. Glasses clinked, people yakked, newspapers and magazines opened and closed. A mysterious alter ego spoke again: You can do this. I asked more questions. She doled out heaping doses of reality. "We don't have the money to do the things we need to do. Small businesses, for example, around Diamond Lake Road and Lake Street are losing business. Things I could do," she said, "is ask people how funds have helped them. Were they disillusioned? Track some of the major tragedies, ask were the funds doing what they were intended to do? If we can get a fund here, will any fund really cover this? With the thirty people I know, map their location across the state. In Minneapolis we have five votes that would support helping the victims. We need forty-seven more."

My palm grew hot and tired from all the writing. I nodded repeatedly as I circled long crooked notes and drew lines to section off more space. I hadn't brought enough paper, so my page soon turned into a jumble of chicken scratch and sideways sentences.

She continued. "You need to convince the suburban and rural senators: their constituents think that we, in Minneapolis, aren't their priority. They don't live here. It's not their bridge. People outside of the city don't care. You have to demonstrate that survivors and those affected are from all over the state." This idea lit a fire within, and I felt a growing excitement that I didn't speak of yet. The overwhelmed pit languished there still, but sitting with her, I already began to dream, to visualize the potential of my involvement. My skills as a technical writer could come in handy with mapping the survivors, with developing a vision and telling a story. Privacy was an integral part of the support group and the collapse experience, so while I avoided identifying survivors by name (if I wanted to name anyone, I'd have to get permission), my brain drew imaginary lines between the software manuals and technical communications that I write at work to this extraordinary predicament. The idea stunned me that maybe this was why I had survived. Maybe even this was why I'd been on the bridge.

"Yes." I nodded. Patrícia continued. The survivors and families need to be more visible. I told her I was surprised—I thought we had been in the media often. She said, "No, you need to be bigger, and the best way is for every survivor to meet his or her senator and make the compelling case." My shoulders slumped. How in the world would we get every survivor to do this? One step at a time, I reminded myself. She warned me that it would be easy to get distracted. "Don't," she encouraged. "Keep clear on the vision and your main goal." Dishes clinked as servers bused them to and fro. Coffee machines revved as I shook my head and asked, "How do we do this?" She said getting bipartisan support would be the goal. Go after the big prize first.

72

I go home and spend the weekend creating step 1, building on that burning idea—an impact map. I print a map of the districts in Minnesota, hunt down someone in the police department who can help me get addresses, and type each survivor's address into the legislative website, the Minnesota district finder, and then I plop a red dot on the map. Once I finish this, I start building a digital version. It takes several days, but the finished map speaks a thousand words. At last. A single picture that shows me what I'm fighting for. We're from all over our state, our Land of 10,000 Lakes and beyond, from the wine country of California to cheese-producing Wisconsin to steer-rasslin' Texas. Just a Minneapolis issue, my ass. I go to work happy for the first time in months. I'm on a mission.

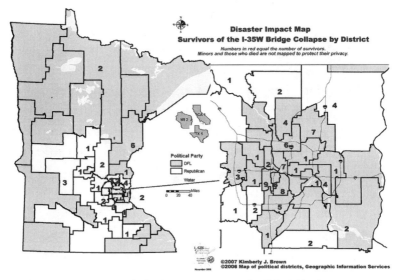

10. Impact map. © 2007 Kimberly J. Brown; map of political districts © 2006 Geographic Information Services.

73

Eighth Fracture Critical Bridge Inspection: September 2001

At panel point #8, stringer #2 has loose bolts, and the bearing block has rotated.

74

I spent all day at the Capitol listening to the legislators discuss the possibility of a victim's compensation fund. They are trying to figure out the legal ways to best set up and dispense a fund. It was a good day but exhausting. My mind was packed full of information.

At the Minnesota State Capitol the hallway echoed the footsteps of people in dress shoes who pushed carts of papers to various hearing rooms. When survivors gathered, those who were widowed wore their dead spouses' wedding rings on simple gold or silver chains around their neck. Seeing them brought back my words with Charlie Gibson at the park, my survival words—*I'm so grateful to be alive*—which now filled me with anxiety. I spent so much time at the Capitol that after the eleventh or twelfth visit, I stopped keeping track. The regal marble floors and hallways began to feel like my workplace. My shoes clicked down the hall in a terse rhythm as I rushed to arrive on time and find the right room, looking for 5 State Office Building, 100 State Office Building, or Room 123 Capitol. Legislators sat at oversized tables behind lettered name plaques. Those months were spent sitting in uncomfortable chairs, and my back ached, but I ignored it as survivors arrived in hard turtleback casts, crutches, wheelchairs, with a world of invisible injuries.

Even though this bridge was their responsibility, the state's, we were going to have a long road ahead of us, our attorneys cautioned. The hearings would come to order. Pen in hand, handouts in my lap, I listened attentively to experts and scrawled copious notes. I used almost all of my vacation time and frequently adjusted my work schedule. Jennifer, my boss, was endlessly supportive as long as I finished my work, which I did. She seemed to realize the greater history I wanted to influence. When I emailed her to tell her of yet another schedule change, she replied simply, "Go get 'em." I stood in my cubicle at the ergonomic desk with tears in my eyes. I couldn't stop fighting. I was lucky to have that support and flexibility. Not all survivors were so fortunate.

Over the winter freezing wind pelted me as I climbed marble stairs from one hearing to the next. I usually came to hearings from work. After finding an open meter or a free spot on the street farther away in the adjacent Frogtown neighborhood, I lugged my work laptop in my generous-sized backpack and showed up in the hearing rooms, my face red from the cold. I sat with other survivors, taking off loud crinkling winter jackets still frozen from the wind. Each time I'd think about ice: a new threat in bridge survivors' post–back injury reality. I now walked carefully, grasping thick silver railings, gingerly navigating broad expanses of government grounds, each time thinking how surreal it was that I was still alive.

Meanwhile, media flanked the hearing rooms. I listened to the drone, the sober chorus of legal lingo like *litigation, liability, indemnification, precedent, claims*, and *tort*—the latter making me and another survivor hungry for pie. My face flared warm when I became aware of camera lenses, when they spun around to focus on me and the others. Sometimes they panned the room, stopping at our faces, and the feeling was strange, like being spied upon. This was a public hearing room, but they were discussing our well-being, our futures. The cameras were like a person who accidentally barges in on you by pushing open the bathroom stall door.

I felt self-conscious and oddly vulnerable. But I was learning, and learning fast, about being with the media. And my confidence grew with practice. Although I'm a writer who avoids the spotlight, I was with the media frequently. So I started applying my writing skills, organizing my thoughts so I could be on point and calm my nerves. Rachel worked with me. Amy helped me form a new paradigm about cameras and reporters, coping devices I would use to channel energy in a more comfortable way. She encouraged me to think about nerves as ego, my own fears of looking bad or stumbling over words. It became bigger than me. It became not about me. I began speaking for those who no longer could. In this new light I donned a brave self. I stood before cameras, alone or with others; I spoke slow and sure with a clear message, in memory of those I never knew. I became obsessed with getting a semblance of justice.

75

The Joint House-Senate Subcommittee on Claims—a committee that hears legal cases against the state—convened to hear a special and important guest: Kenneth Feinberg, the special master in charge of designing, implementing, and administering the federal September 11 Victim Compensation Fund, a federal statute signed into law by President George W. Bush. Appointed by the United States attorney general, Mr. Feinberg shared what he had learned during his extraordinary role in history. This hearing took place on November 10 at the State Capitol.

Mr. Feinberg—a tall Caucasian man who carried himself with an authority beyond his receding hairline and sober gray suit—described the process of discharging unlimited recovery funds in the aftermath of the 9/11 attacks. He spoke of dispensing the fund. It was an unprecedented action to an unprecedented historical event. There were many reasons for it—one being to assist over two thousand dead and injured claimants and take their recovery suits out of the tort system, which could threaten the entire airline industry. To a Minnesota legislator's question of whether the state had to determine fault as to why the bridge collapsed, Feinberg expounded on his experience, giving us in Minnesota much to ponder and much to guide our decision. But the remarks that caught my ear were these: *If a fund was the route they chose, timeliness was critical. Victims are distraught, grieving, and angry.* He explained in layman's terms that the verbiage of the law tends toward the clinical and that we should keep that in mind. The 9/11 Fund made "no attempt to place a value of the worth of your dead family member. It only sought to do what judges and juries do every day in America." He sounded every bit the lawyer as he continued. Recognize that while frustrating, money is a second-rate way to help, it can't bring back loved ones. He urged: "Simplicity is a virtue. Don't make these people wait. Don't hold them hostage." He said the 9/11 Fund was the "right thing to do" to demonstrate the national will of the American people.

Feinberg stressed the importance during the claims process that every 9/11 claimant had an opportunity to be heard. Each person came to explain his or her situation before a live person, under oath, and with a transcript. Thirty minutes apiece, Feinberg met and listened to 1,500 people who needed to, in his words, express their views and their concerns with how to handle the future (medical needs, as one example). Feinberg said this process was important to "allow people to vent life's unfairness." I was taking notes, like usual.

A fellow collapse survivor wrote on the edge of my paper: "I would have hated to be the 1,500th person. He was probably so numb by then."

The hearing lasted for hours.

We heard testimony by Peter S. Wattson, Senate counsel, State of Minnesota, based on his handout "Compensating Victims of Bridge Collapses Outside Minnesota." He spoke of historical bridge failures and the compensation that was or wasn't offered. A bridge in Tennessee with pilings made of wood that should have been on dry land but wasn't. "Rip wrap" that was supposed to be put below surface was not. "Failure to maintain it around the bridge piers." The river moved the pilings. That state concluded with no judgment; instead, it was a settlement. New York doesn't have a liability limit. Payments done through operating budget (tolls and such). More successful because not only liability from errors in design but also a fraud element. The double-decker San Francisco–Oakland Bay Bridge, whose piers were built to withstand earthquakes but didn't. Twenty days after the failure, California's governor signed a bill to compensate victims in the form of a $30 million appropriation. Money was made available immediately while claims were pending. A Red River Pedestrian Bridge in Arkansas, dubbed the "swinging bridge" because pedestrians could *move it*, was closed to automobiles in 1972 but later fell with between thirty to fifty people on it. The bridge had been owned by the county, and the state claimed constitutional immunity. A bridge on I-40 in Oklahoma that was struck by a barge. State limits were in place: $1 million per incident, or $175,000 per person. Legal suits followed against the barge company. Many other bridges have fallen after being compromised by an outside influence and so on. Wattson also spoke about differences between states, the limitations of the law in rating people on value of a life, earning potential.

After lunch the meeting continued with testimony from survivors who spoke of their concerns for the future, contrasted against their injuries and the trauma they'd been through. Blacking out, flying through the air, cheek reconstructed with three pins holding it together, brain surgery, heart stopped for three minutes, CPR to revive, lung punctured from the CPR, fractured ribs, broken backs, herniated discs, PT, OT, bills, can't work because of injuries, decimated savings.

As the long afternoon continued, more experts testified. We listened to what felt like endless testimony about no-fault automobile insurance law, workers' compensation, other cases in which individuals brought claims against states, other states' caps (ceilings or limits for each person's compensation), the reason for caps—to provide financial stability, for starters—and the way other funds were financed (Virginia Tech, e.g., composed entirely of private contributions).

A man named Al Gilbert testified regarding the concept of "tort immunity"—originated in 1788 involving a claim that a county had failed to maintain a bridge! A court determined that you're suing the king. It originated from a concept called "sovereign immunity" that in its literal translation means that the government can't be sued without its consent. The Minnesota Supreme Court determined in 1962 that it's no longer appropriate for municipalities, but they kept it for the state. In 1970 they looked at it again—there was a Minnesota Supreme Court case. In 1975 sovereign immunity was again abolished for municipalities.

The government has unique responsibilities to the public (e.g., snowplowing, public policy, taxes, welfare payments). Reasons for the caps include a concern for financial stability—for instance, to prevent states from being viewed as having unlimited funds. Not having caps could chill the efforts of employees, who would worry, "If I do this, someone will sue me." Caps give more latitude, and the issue of "deep pockets" is not upheld as a rationale. Gilbert discussed suits against other states and how they were handled. A public policy question: should you wait to determine more certainty to wait for cause? Gilbert thought Minnesota could move ahead by a fault or no-fault system; California seemed like a good model; the Emergency

Fund got money out quickly in the short-term, which could be offset later if the reward was decided later.

The State of Minnesota has a cap on how much it can be sued. At the time of the collapse, these amounts were $300,000 per person and $1 million for a single event. This would be a problem. The fund could be a way to help all of us, in a reasonable way. MN/DOT fired the emergency response manager, who failed to return to the state for ten days after the collapse. The 35W Bridge was owned, inspected, and controlled by MN/DOT at the time of the collapse. This happened on the state's watch. It wasn't terrorism. Wasn't a force of nature. It was a 100 percent preventable disaster.

When it was finally time to go home, I drove with Mr. Feinberg's words in the forefront of my mind—especially "Bad things happen to good people more often than I thought." It had been a long hearing and an unwieldy hunk of tragedy to digest.

76

The next hearing, with the House State Government Finance Committee, focuses less on whether to help but more on how.

State representative Ryan Winkler's proposed legislation to create a survivor fund moves forward, officially House File #2553. They replay video from Feinberg's testimony and continue discussions. From attending hearings, I'm beginning to recognize faces and learn names: Representative Ryan Winkler, Representative Chris DeLaforest, Representative Melissa Hortman, Representative Bill Hilte, Representative Phyllis Kahn, and others. The highlight of this hearing: the notion that we don't want to have to put up signs that say, "Cross this bridge at your own risk."

77

Ninth Fracture Critical Bridge Inspection: May 2002

At panel point #8, stringer #2 has loose bolts, and the bearing block has rotated.

78

On January 15, 2003, Lieutenant Governor Carol Molnau, also commissioner of Minnesota's Department of Transportation, invited Minnesotans—taxpayers and state employees—to submit cost-saving ideas to the MN/DOT website (news release). That year she also directed MN/DOT managers to submit a list of budget-saving activities before the end of January. "Our success depends upon our ability to work together to find new ways to cut costs and make MN/DOT work better," Molnau said. "While we face many challenges, we can accomplish this goal by remaining committed to our core business—improving the state's transportation infrastructure."

79

Following hearings, the media reported the progress and issues that the legislature faced. I composed emails to the committee members to weigh in, and both behind the scenes and in person at meetings, I stayed immersed in the process. At the conclusion of one hearing, as people were filing out, I met one of Minnesota's senators—Senator Jim Carlson. He opened a three-ring binder and showed me page after page of inspection reports from the 35W Bridge, specifically a broken bolt and a spacer that had rotated, which hadn't been repaired. Year after year it was photographed and logged but ignored.

When I got home from another hearing, a letter had arrived in the mail from Senator Torres Ray. I sliced it open with a letter opener and read her handwritten note. "Thank you so much for taking the time to meet with me. I admire your strength and your concern for others." Reading her words reminded me to not lose hope. I admired the state's seal on the stationery and placed the card on the kitchen table.

But the hearings took much out of me. At work it was tough to concentrate. I felt antisocial and just wanted to go home and be alone. Exhausted and depressed by the enormity of it all, the road to a solution felt endless. At lunch with coworkers, I listened as they talked about high-speed Internet access, modems, and credit card interfaces. Others commented on how good someone's stir fry smelled or how lame someone's sandwich looked, and I stared out the window, all the legalities and arguments weighing on me and dominating my thoughts.

Later a survivor friend broke this heaviness momentarily. A vegetarian—she emailed me a picture of two hot meat sandwiches. She noted, "Don't they look appealing? Not." The picture blurry, the layers of meat looked grotesque, stuffed into a processed white bread bun. I held to this tiny moment of levity. We revisited the idea of the meat sandwiches during long hearings. I'd say, "I could really go for a hot dago right about now," and pat my tummy like it would be just the ticket. I added news about my height. At five feet two and one quarter, I wasn't tall to begin with. Three-quarters of an inch shorter from the collapse, I joked that although I'm shorter now, I'm louder. Eyes tired, back aching, hungry, and my mind drifting, I thought, what a relief it would be to go home to Rachel and our zoo—Lucy and our three cats. Right after this thought, I remembered those who couldn't go home to their partners anymore. I stuffed the handouts in my big backpack and bundled up against the cold.

80

Tenth Fracture Critical Bridge Inspection: June 2003

Panel point #8—"one bolt broken off at south floorbeam connection. . . . Stringer #2 (south side): one bolt is missing and the nut is missing from the other bolt. The bearing block has rotated 90°. [1999] Missing bolt replaced. [2000] Bolts are loose, needs repair."

81

More all-day hearings. Before the Joint House-Senate Subcommittee can get started, we sit through other claims. In government they set a time, but when you actually get there, start times can be changed, delayed, or otherwise held up by someone not showing up or who else knows why. That day an inmate lost his medicine bag during his incarceration. His testimony sheds light on what this bag means spiritually and in a religious sense in

Stringer 2 Bearing Block Rotated

11. Bearing block still rotated. Bridge 9340, Fracture Critical Bridge Inspection Report, June 2003. Minnesota Department of Transportation.

his culture. In the bag, lost by the state, was his umbilical cord. No matter, it feels like forever before they get to the bridge collapse.

When they finally begin, Representatives Diane Loeffler, Tom Anzelc, and Dennis Ozment and Senators Ron Latz, David Hann, and Rick Olseen call the hearing to order and begin discussing a proposal for an emergency hardship relief fund. This bill would be an interim measure, attempting to address one of the short-term losses. We would continue working toward the greater bridge bill after the holidays. We hear testimony from people like Tom Chuckel with the Department of Administration; Peter Wattson, Senate counsel; and Kathy Pontius, who speaks about subrogation—the ability for a company to "stand in the shoes" of the patient.

I write on my notes, "My meter runs out at 2:25 p.m.," and show it to the survivor sitting next to me. I write in the margins, between paragraphs, and on the reverse side of handouts titled "I35W Bridge Claim Summary," "Concepts for Interim Claims Relief," "Risk Management Claims Department," "Compensating Victims of I-35W Bridge Collapse," "Use of Trunk Highway

Funds to Pay Tort Claims," and so on. I write, "It's freezing outside," and underline it several times.

Emergency relief legislation passed at the Capitol in time for the holidays. Survivors who had lost wages as a result of injury or death would be eligible to tap into a $1 million fund, up to $10,000 per claimant. From what people told me about government and how slow progress moves, this was a small miracle.

I should be happy. I'm probably just tired, but I don't feel like doing anything. I have no energy. Trudging. I don't want to do lists, to-dos, or groceries. I don't want to think. My body feels bad, mired. I have lost the joy for cooking and writing. I'm going to go to bed and forget about this day. There isn't enough time for everything. I hunger for beauty, but right now the view outside is a muted gray. Light fades early in December in the Midwest. Salt and dirt cover everything, and temperatures barely top 7 degrees. A bridge falling now would throw people onto the ice. I wouldn't want to fall today.

New reality: I think as I drive, "Careful, careful, just get across this bridge." I try to convince myself that I'll feel more like myself someday, but I'm still back in August. The weather should be scorching. I should be playing soccer, running, feeling good, making friends. How far from all of that I've come. I think about working out, and it feels too hard. I put off going downstairs to the treadmill. Instead, I clean, reorganize the shelves. I start to know, inside, that I should try more therapy—individual therapy, not the support group. I went to Kris for a few sessions, covered by my employer. I need to go back. I will, when things slow down, I tell myself. But deep down I know: I can't keep distracting myself. I have to start dealing with my crap.

82

Eleventh Fracture Critical Bridge Inspection: June 2004

At panel point #8, stringer #2 has loose bolts, and the bearing block has rotated.

83

The first snowstorm on December first. Mentally, I'm still in summer, in the month when time stopped. Trying to beat the snow, I rush to get chores done. Garbage, recycling, dog duty. Exiting the alleys of south Minneapolis with Lucy, the pocked snowbanks resemble coral. When we left, the snow started falling and worsened, blowing into the gaps in my hood and scarf. I yelled to Lucy, who was at the end of the sixteen-foot retractable leash, "Come back." I walked methodically and purposefully and stopped on ice, worried she'd pull me over or that I'd slip. Snow stopped us in our tracks. Maybe the snow is here to stay, so we take the time to notice, appreciate nature, quit complaining, and use this time to appreciate our lives and loves.

At group, survivors talk about how their lives have changed, how they take the good with the bad. Most of them young, they hadn't imagined worrying about ice being dangerous at their age, at the start of their adult lives.

Back home, under a toasty blanket on the sofa with a cup of tea, a curtain of snow falls in a tranquil hush. Next summer should be sweeter, for all we've seen this year. Dinner's piquant fragrance drifts through the house. It'll be ready soon: a quiche in the oven and a slice of chocolate pie to split. Rachel and I are both here, together still. I feel joyous and guilty about that all at once. We'll never recoup time, but we measure these days with a different ruler.

During another December snowstorm I spend two and a half hours in traffic, cars and trucks crawling. All I can think is, "Thank god I'm not on a bridge." On the news people talk about their long commutes—routes that usually take thirty minutes taking two or three hours, cars towed, gridlock. In one interview a man said of his commute, "Today is not a good day." Almost two hours into mine, I think, "It isn't so bad." I still have my car. I can physically drive myself where I want to go. I think of people in group who cannot do for themselves. Then I go bigger and think about people, not only in this country, who never had the privilege of driving. And lucky me, I have a two-and-a-quarter-hour commute home, warm in my cocoon.

Most of all, I'm still here to laugh, to listen. I'm finally getting how to deal with traffic jams. I can use the time. I borrowed audiobooks from the library. In the CD player, Amy Sedaris's *I Like You: Hospitality under the Influence*. She talks about cooking, uses bad grammar on purpose, and makes jokes about things that aren't "legal." She says, "Don't put everything on your plate that is white unless your theme is to hold a KKK party." I shriek at the terribleness of the joke and grin out the car window. Puffy marshmallow flakes fall and fade on my windshield.

84

Twelfth Fracture Critical Bridge Inspection: June 2005

At panel point #8, stringer #2 has loose bolts, and the bearing block has rotated.

85

Surrounded by people in wheelchairs, walkers, and back casts, when my back pain worsened, I ignored it. But one afternoon, after catching a movie at a theater, pain brought me to a halt. Hunched, I winced and straightened like an elderly woman. The first week of December, I went to Consulting Radiologists, Ltd., for an MRI.

The technician asked, "Did you have an accident?"

"We fell over one hundred feet. Not your average car accident."

We talked. I answered questions. Did you have surgery? No, just dull aching pain. The MRI machine filled a room. Cartoon fields, flowers, sunlight and clouds, decorated the ceiling. Before turning on the machine, the technician instructed me to lie down, and she handed me a blue rubber ball.

"Squeeze this if you need to stop."

She gave me headphones and then disappeared into a booth behind a wall of glass. A radio station piped into the headphones, and she asked me how loud to adjust the volume.

When the machine started, it thudded for twenty seconds. The loudness startled me. Then it boomed for three minutes. The banging overpowered Cities 97, Dido's song "Thank you."

She sang, "I-I... want to thank you... THUD THUD THUD... [*faintly audible*] for giving me the best day... BOOM BOOM BOOM... [I filled in the chorus] of my life." The testing vibrated beneath my back. (Don't move!) Between thuds and booms, I took shallow breaths. Each time the tech reminded me, "Don't move," I wanted to move so badly! I felt the ridges on the ball. I closed my eyes. Could anyone fall asleep during this test? THUD THUD THUD. No.

Afterward the tech pointed me to the dressing room, where she said I could ditch the hospital gown. I opened the locker where I'd hung my clothes, got dressed, and finished by draping a wispy black and silver scarf around my neck. It had been Mom's. I donned a black peacoat, also Mom's, which oddly enough was by the brand "Karen," Mom's real name. It was a tad too tight, with sleeves too short, and as I finished, my mind flashed on a car accident. I rode in a car and smashed head first into another car, then into something hard like a wall. I raised my hands and arms, gulped, and tried to breathe. Time's up. My eyes darted, and then I died. I imagined my clothes, jewelry, and scarf underwater and wondered what they'd look like after being in the river. Could they be restored? Could someone just wash and re-wear them? I hoisted my bulky backpack onto both shoulders, realizing maybe it wasn't such a good idea to wear her things.

Headed to the parking lot, I left the warm hallway and stepped into the winter afternoon, feet squishing through salt, slush, and snow, MRI films in hand. They didn't X-ray my head. I repeated the acronym for MRI, magnetic resonance imaging, wondering if it would show anything. During the drive home, black birds hovered, swooped, and freckled the sky.

86

Thirteenth Fracture Critical Inspection: June 2006

At panel point #8, stringer #2 has loose bolts, and the bearing block has rotated.

87

To repair me, I must invest in my body: the miraculous system of blood, bones, sinew, head, and heart that survives assaults of all kinds and holds onto hope.

In the dim light of Amy's wellness office, the Dogwood Studio, I lie on the massage table under a cotton sheet and gaze at the metal tiled ceiling. In each section scrolls and turtle shapes repeat. Round paper lanterns hang from long cords, and white seams circle the delicate orbs and bind them together. I think about the mystery of Mom's suicidality. Calmed by these surroundings, I tilt my view skyward and wonder aloud, "When did it start?"

Amy presses into my back. Aided by massage oil, her fingers glide along my spine like a team crewing down a river. I'd never know the truth, couldn't ask her. Mom was gone, but someone had clues. Someone noticed things amiss. Was the door ajar, the mail still in the slot? What taught Mom to hate herself? How had she grown from broad smiles, dressed in a cotton baby gown, with her mom, dad, and doggy bolstering her?

Bodywork involves letting go of intellect. The release into sensations and hidden knowledge the body has known but the mind wasn't ready to comprehend. The plethora of feeling within the vessels of my body's bridges carry all the memories that have shaped my life. Inherent in the invisible structure: tension and compression, load and stress. We downplay their importance. But recognition heals me.

Muscles near my injured disc fire. Amy leans into her forearm against the width of my back like a Zamboni cleaning ice.

She tells me: "You are doing so well, Kimmy. Let the love and strength fill your heart with joy."

Everybody should have beautiful Amys in their lives. I tell her how much I appreciate her and what's been bothering me. I say, "I want to let love and strength fill my heart, but I couldn't stop crying Sunday night." I need to commiserate, and I can't see any alternative to the despair I felt. I ask her to keep happy thoughts for me.

She reminds me what she's been saying all along. Tears aren't bad. They cleanse and help us let go of worries. "Love all the things that are going well. Be joyful in your heart and celebrate all the goodness you have in your life. Once you start looking around, you will see it everywhere." I plodded into the day, her words like earbuds encouraging me. Diabetes. Bridge collapses. Amy's words remind me that I don't have to tackle these macro problems all at once.

Walking to my car, parked in the work surface lot (not in the ramp), I think about how Amy had pointed out the positive. I'm transforming, experiencing another metamorphosis. She's right. If we reject negativity and bitterness, if we're intentional about what we'll allow to influence us, we realize we receive everything we need. When we reach out, open ourselves, look for beauty, and try to put beauty out in the world, we reap what we sow, and mysteriously, we're provided for.

88

Nearly five months after the collapse, Paula Coulter (one of the most severely injured survivors) finally went home from Courage Center—a renowned rehabilitation facility.

Coinciding with this, my back felt improved. I had greater range of motion, and I could reduce chiropractor visits to one per week. The MRI showed borderline degeneration in lumbar disc five (L5), bulging like the smashed filling of an Oreo cookie, white breaching the black. Surgery isn't necessary; continue therapy treatments. On my side, elbow tucked into an *L* by my pillow, I saw a thin cord, a finite two-inch line on my forearm from years ago when one of our cats clawed me. I'm sure I had picked the scab, loving how a whole straight sheath of skin could be removed from the body, leaving in its wake a smooth surface. Faint scar, barely visible, evidence.

On the drive to work, on the outskirts of downtown Minneapolis, steam rose in a voluptuous plume. One below zero. I flashed to the opposite, to catastrophic black smoke. In my squished disc, I felt the road bump. How did a fellow survivor "relax" (as she described it) as she fell with the bridge? Crossing an overpass, I saw the road split open and heard the sounds of falling, breaking, and my own screams. Then I imagined relaxing. Or trying to. Even pretending was unsettling. Could I ever do this? Should it be a goal? A semitruck with hardened snow on its roof shed pieces. They sailed through the air and rained on cars. A chunk abruptly lifted and flew through the air. In slower traffic, my god, I realized that people parked underneath Highway 394. The entrance and exit paths in and out of downtown Minneapolis are elevated like an elongated bridge. All this traffic passing overhead? I wanted to tell those people, You could be crushed.

When Rachel was out of town, on the phone she said, "I miss you so much, it hurts." "Me too," I said. My coping mechanism: I thought about ourselves as lucky. We weren't like the bridge widows. She would come home. I wouldn't count my chickens yet, but I would get her back.

Another session, with tiny glass jars that look like empty candleholders, Amy cupped my back. She lit on fire what looked like a big cotton swab and then held the flame in a cup. She removed the fiery swab, blew it out with a big breath, and then quickly put the cup on my back. Immediately, it sucked my skin like a baby with a pacifier.

Cupping techniques date back to AD 281 and are based on the belief that a person's energy, or qi (pronounced "chee"), can be blocked along pathways or meridians in the body—a basic concept in Chinese medicine. Qi imbalance can be caused by blocks due to external or internal stress, lack of sleep, or overconsumption of alcohol. To restore balance between the mind, body, and spirit, cupping is just one modality that can be used.

Amy moved the cups around. Massage oil allowed them to slip stiffly across my back's landscape, which was like an oil spill—like the extra virgin olive oil I streamed in circles in a frying pan when I stir-fried onions and peppers.

Not left to settle, cups sucked the pain from my body, broke up scar tissue—disturbed it the way sorority and fraternity parties disturb a campus, music bumping, people pouring onto lawns, kids inebriated, making new connections—my back's version of a kegger.

The marks proved it. Cupping left perfectly round purple circles. Showing Rachel, I lifted the back of my shirt. "Oh my god! What did she do to you?" With a wink I explained in a funny voice that it's a holistic practice based on an "ancient Chinese secret," not even realizing my cultural reference to the commercial from the 1970s for Calgon water softener.

We want our bodies to behave. Yield to our will. Bend when we say bend and straighten when we say straighten. But we need to listen to our bodies, not the other way around. The body behaves as if it had a mind all its own. My true mind, my brain—a pathetic, out-of-shape, bench-warming player in this whole drama—receives commands. I send messages to my muscles, and for a spit of time the muscles obey. Gretel Ehrlich, author of *A Match to the Heart* and a woman who survived being struck by lightning, says she "had a lot to learn about getting well." True for me too. Amy tells me that bodywork will help prevent emotions from becoming "stuck." Ehrlich says the body has a hundred billion nerve cells. "Intelligence exists everywhere in the body, not just in the brain." Science tells us that trauma can physically alter our bodies at the cellular level.

I want to heal. Practitioners stretch my muscles and crack my bones. I push water and avoid alcohol. Sensitive, my mood shifts like a breeze through an open window—a curtain blowing, wind chimes ringing. I tend to my body like a gardener tends a plot of earth—planting marigolds and hostas, running a red enamel spade through a rich layer of soil. Like old friends, hostas return each spring. Marigolds stand bright in the summer sun, but they won't return next year unless someone coaxes the seeds until they blossom again. Pulling apart, an injured disc in my lumbar spine holds onto the bridge fall. My back is a marigold. Like the ephemeral nature of an annual, I tend, pull loose, and reseed the life cycle of health.

89

And yet a competing side of yourself slides toward numbness, instant gratification, comfort food. That side has no energy left to grocery shop, plan, or cook. You look back on how you've spent life up until that fateful moment, and you realize too succinctly that the amount of time we're given is finite. You don't know why you would choose to spend time doing mundane errands or chores. And there are other things. People who don't know you see a woman who looks fine on the outside, and right now that bugs you.

You are drawn to food like a prisoner sitting down to a last meal.

Croissants, bakery fresh bread, macaroni and cheese, California burgers, potato chips, French fries, tater tots, the speed and ease of fast food, bean burritos, tacos, mashed potatoes, fried chicken, Taco Bell. You administer two custard-filled chocolate éclairs to each hemorrhaging ventricle of your heart. Defenses worn down, you talk to yourself like a bully. You're dumb. You can no longer do ordinary things like ride an elevator or park in a parking ramp or fly in an airplane. Maybe tonight you'll order a pizza or get a bunch of burritos or tacos at the drive-through. Yeah, that sounds good. They soothe you from the inside and blunt fear that ripples through your post-traumatic reality.

Post-collapse, you will gain all the weight back. The word *moderation* has been removed from your vocabulary. You cannot get enough. There is too much bounty to be had. This isn't helpful to you, not health wise or emotionally or anything. But it is what it is.

The trauma survivor wakes each morning and thinks, "I lived another day." We look for ways to soothe ourselves or celebrate. For others their script might vary in any of the following ways. What can I eat, drink, smoke, shoot, imbibe, do, that will numb the pain? Except it's not that obvious, the thinking. It's more like: "Wow! Another morning, still alive." Make it a good day. Rewards. Happiness. Instant gratification. Because knowing that life is short, toiling becomes less appealing. Why spend it on less-than-satisfying things? Luxuriant smells, marvelous sights, satisfying textures, safety, warmth. These are the goals. What gives all these things? An egg and cheese croissant. The butter. The savory flakiness. Even better, toast it on the grill to crisp the edges. Paper thin layers melt on contact with taste buds. The French eat them all the time, and look how slim they are. Yes, I'll be French. For whatever time remains, savor life now.

90

For Christmas my friend gave me peppermint body wash—my favorite scent. Knock on wood, the hives are gone, so with near giddiness I pour crisscrossing lines of peppermint on the scrubby. Running it over my arms, legs, belly, I examine my body—amazed it's so intact. Amazed that I walked on two legs away from that. But also paranoid that somehow my body will fade, slowly or quickly—that one day I might wake to find arms or legs that I scrubbed yesterday are no longer mine. Something might happen, maybe an accident, and slowly I might have to say good-bye to my health. And yet, why do I go here?

For the thirteen who fell in the water and couldn't escape, I say prayers. "Why pray for them—they're gone? Pray for people who are alive," some people might say. Yes, but those people could've been me. Every time I'm in stop-and-go traffic or I stand in an elevator that's too full, I guess at the panic they felt. Can't breathe. Can't move. Can't forget.

91

Publicized in the local *Star Tribune* newspaper article "Phone Call Put Brakes on Bridge Repair," by Tony Kennedy and Paul McEnroe (August 18, 2007), I continued reading about the bridge history. During the latter half of 2006 the Minnesota Department of Transportation was in discussion with a contractor (URS) prepping to begin a reinforcement project. Records show it was debated whether to reinforce, reinforce and inspect, or inspect only.

"Internal MnDOT documents reviewed by the *Star Tribune* reveal that last year bridge officials talked openly about the possibility of the bridge collapsing—and worried that it might have to be condemned.... The documents provide the first look inside MnDOT's decision-making process as engineers weighed benefits and risks, wrestling with options to prevent what they believed was a remote but real possibility of the eight-lane freeway bridge failing."

Owners of land beneath the bridge and the Coast Guard would be notified of the work, and new weight restrictions would be enacted in preparation for addressing the bridge's glaring weaknesses. Experts warned that the bridge, if not reinforced immediately, could collapse but that the drilling had the potential to weaken the structure. On January 17, 2007, MN/DOT about-faced, abandoning plans to reinforce gusset plates and steel, and "embraced inspections over reinforcements" and decided to do nothing.

92

After catastrophe, everyone thinks, "You're safe now." End of story. I tell myself I'm fine. But I'm so afraid. Of being lost. Of dying. Of pain. Of being separated from all I love. This is the paradox I imagine many survivors feel. We've glimpsed the potential. Seen, imagined, dreamed, felt, and smelled the acrid burn, twisted wreckage, the sheer cliff of the end of life. Someone spoiled the ending to a movie. Now I know. We know. Something bigger is coming. Will we have time to mourn? To let go? To struggle against fate? Or will the end sweep us away like a reef crushed beneath a tidal wave?

The not knowing is hard. Fuels my fears. I don't want to ever let go. But someday, someone or something will decide for me. I suppose this will be the psychic work that I will face in the years to come. Learning how to let go. Learning acceptance. Someday this will all be gone. These anxious moments are where the old splits into the new, where the end changes from a vague faraway idea to a cosmic knowing. A sense that nothing is sure in life. Nothing. And am I ready? What is the point to any of it? I'm still trying to figure it out. I'm full of some answers, entirely my own, but it's going to take more time, which luckily I have, but how much?

So, interruptions in the routine bring feelings to the fore. An unsettled sense of placement, of time. I need to take a brief road trip to Omaha for work, but I'm afraid—although I know I can't live that way. Yet the truth isn't dimmed by this determination. I must get home as quickly as possible. I'll bide my time, looking at the clock, calculating how much longer. For now we're all here. Everything I think, feel, and know makes even a day away hard. My center is here, only here.

When people ask: "Don't you figure you survived this terrible thing—you must have important things yet to do? Don't you feel protected, like bad things can't happen to you?"; I respond: "Those thirteen people had important things to do.... Who's anyone to say that it ended at the right time?"

For me phrases like "It was their time," "God wanted them back," point more to our own good-intentioned desire to make sense of the unbearable. What if fate was wrong? What if it screwed up and one of the thirteen was supposed to be here—instead of me?

To this Rachel said: "When you feel like you don't deserve to be alive, it's hurtful to me. Because I need you. I need you to be alive. You're no good to anyone if you're dead. I deserve to have you be alive. You're supposed to be here. You were supposed to survive that catastrophe."

In the morning cold of the car, my glasses fogged. Odometer start: 39,764. Backing out of my driveway, sadness emanated through me like frost, spread stealthily like a rash. Five a.m. As soon as I backed out of the garage into the driveway, I sobbed, gazing at our brick one-story house, light illuminating the windows like eyes. I mopped my tears with a napkin and realized a simple truth. Home's the one place I feel safe, where I've begun to recover a sense of wholeness. Coming home after a trip will feel like it did the night I survived.

A pang ran ribbons over memory.

I pulled away and steered the dark alley into the cavernous mouth of morning. My eyes itched from crying. The inky night seeped in and wrapped around me. I followed a pickup truck, and its headlights led the way. Then the brake lights brightened, seemingly for no reason. I followed suit, dropping my speed by ten miles per hour. What had the driver seen? Why had the driver slowed? I could see slightly around the truck, so I peered ahead to the left and right of the blacktop. Nothing. Cars passed on the opposite side, and I favored the shoulder when a realization surged. What if there's nothing out here to protect me? Bad things, tragedy, can happen at any time to anyone. Can my slowing, my watchfulness on a country road, alter the trajectory of what's bound to occur? How do we live, knowing it won't last, guaranteed?

I used to view the open road, the expanses of earth and undulating hills and outcroppings of trees, and I'd sigh—the land before me signaling freedom and possibility. Not today. Today I felt hurt, loss. I couldn't shake the dread of driving into an oblivion I couldn't prevent. The road represented an unpredictable chasm. Sun crept up a bank of trees, fiery orange, like grief that roared behind my eyes. Lonely is the road. The crying urge pulsed and cycled. I drove farther and farther from home. Soon the odometer would roll to four zero zero zero zero.

The families of those killed felt, sometimes, like their loved ones weren't theirs anymore, like the public had taken them. We wanted to sympathize, honor them, remember them, say their names—a natural impulse. Rachel would've gone through what they have. But the law would have refused to treat her with the acceptance and love I would have demanded for her. They'd call her "a friend." She'd be reduced to a loose title that anyone on Facebook could be named with a click of a mouse.

I think of the people who won't get cards anymore; think of the thirteen deaths and remember how the consequences of death aren't calculated by simple math but by ratios. That one person vanishing changes the trajectory and fate of—could be any number, but let's say 13—13 lives around them. It's one reason why the "death toll" is so ridiculously ill named. Apply this rough ratio, and there are 169 people whose 13 might as well have gone to Mars and never came back.

When I was young enough to voice my questions aloud, Mom confirmed, Yes, we'll all die someday.

"I'm gonna die?"

"Yes, but not for a long time."

"How long?"

"Hopefully not for a long, long time. When you're really old. Older than Grandma."

"How old is Grandma?"

"You've got lots of time. Now go play."

See? This is the fallacy that makes sudden and traumatic events so unbearable. Ripped from each other, your baby with her buttery voice gone . . . it's backward. Nobody goes into business thinking: "It won't happen. Not for a long time." Insolvency, bankruptcy, failure, the end. Businesses plan for it. Prevent it. That's the challenge and gift that I struggle to fully comprehend and act upon. Because if I know, if I truly accept, that yes the end will come and not just as a general fact like how gravity keeps us from hurtling into the sky—if I not only know but feel it, could the arc of my life be different?

This fact nags me, prevents a simple trip from being just a trip.

A risk. Dangerous. Thirty thousand feet of physics. A steel tube stuffed with people, suitcases, and beverage carts lurched through the air at hundreds of miles per hour. We take modern flight for granted as I take for granted Rachel's slumber by my side, that tomorrow life will continue. Knowing this truth—that I'll die, that Rachel will die—affords me the most power I've ever held in my puny hands. I start counting time. Every day, even Mondays. Nothing measures too minute to lift and weigh. What do I want to do before it ends? What do I want to be?

Homeward bound, the work trip finished. Leaving Nebraska, I crossed Nebraska's Mormon Bridge, a double bridge like 35W was. The enormous steel frame hulked over my head like an Erector set, and memory rushed forth. 35W's trusses were a frame below—those broken bolts had stayed broken. Would this bridge have better luck, everything visible high overhead? I imagine what it'd be like to fall this winter day, trusses topping us. Black river below. A barge and city lights in the distance like outer space.

My thoughts returned to familiar places. Cleaning the house before I left. With a new vacuum, I had extended the arm, ran the brush across the dust lit by the glow coming in the window. I stripped the bed, cleaned it from pillow to foot. Moved the dresser. Sucked up hair balls. After that road fell, I thought there was no sense in any of it. Why go on? Why try? Why invest in this house, this place where we sleep, us? Because at night I'd turn the knobs on the lamps, just dusted. I'd see the room glow blueberry, the color of our walls that we selected together, and I'd snuggle in, smell soap, Rachel's shampooed head near mine, touch her hand and pull her close after kisses good night. I'd invest in this room for its potential.

I cut across Iowa on Interstate 80 East. A gaggle of semitrucks flanked the highway, dotted lights outlining their boxed trailers. The sky lightened the longer I drove, high speed, through the oblivion of early morning. I couldn't see outward, just a halo of headlights catching in front of me. Another hour, and the straw-colored sun would peek over the horizon. A few more miles, and I'll lift from a valley where fog thickened over hoarfrosted fields. The drive will turn cement gray, then salt white, and it'll all be easier.

Home! I slid the key into the lock, grasped the smooth doorknob, turned it, entered. I heard the lull before Lucy woke from slumber. Then a racket,

Lucy's claws clambered across hardwood floors, as she took the corner by the upright piano, her keen voice yelping a high-pitched greeting. She wiggled like a noodle, and I petted her as all the familiar colors, smells, and sounds of home returned, and I planted a big smackeroo on Rachel's lips.

At day's end I pulled the line of Rachel's arm within mine. I wanted to build a long list of nights simply cuddling, her body like that pickup's headlights shining gauzily through the unseen. I wanted to follow her on that road and imagine that it never ends but in dreams where we wind up in the same car.

93

People didn't attend the support group only to help themselves. They were there to share stories, get support and give support, and generally to just feel validated that they weren't alone. People were sometimes vulnerable; some just wanted to talk; others wanted and needed to feel angry.

Yet the word *survivor* was supposed to encapsulate an identity, but it was a catchall label. Pick random people. Some are nice, some are funny . . . some are not nice at all. The people in the group spanned the gamut. I want to say we all got along to affirm the sense of togetherness reported in the media, like we were a homogeneous and emotionally close-knit group. For some maybe that was true. But in any group randomly chosen, some bonds developed, and others did not. As support group meetings continued, it grew clearer. We were a misshapen bunch, strangers thrown together by chance with only one experience in common. United by a thirteen-second fall.

We were in contact with each other via an email list that continually grew, and not everyone played by the same rules. The range of personalities, senses of humor, and backgrounds became apparent in the natural cliques that formed and in the conflicts that erupted. For example, one survivor forwarded a slew of sexually explicit jokes. One joke showed an animated cartoon of a woman's bikini-clad bosom heaving in a lewd display replete with sound effects. Content aside, the file size alone clogged people's in-boxes, in some cases blocking business emails. People were understandably upset, and the sender apologized. We gave him the benefit of the doubt, chalked it up to bad judgment, and new ground rules were set.

The group was supposed to be a safe place where we could connect and feel united in our experience. I can't speak for others due to confidentiality, but for me there were circumstances that sullied the experience. It only takes one bad apple, and this cliché proved true. I fended off unwelcome sexual advances, occasional ignorance, and homophobia regarding my relationship and marriage. Once, when I introduced Rachel as my wife, a fellow survivor laughed like we were ridiculous. Going to group and navigating the vast personalities was a source of friction and extra stress. I didn't particularly gel with some people, and some didn't take a shine to me. But I did my best to dwell on the positive. We weren't a perfect group, but there was power in numbers.

94

August 1, 2007

The I-35W Bridge collapses. Of the 180 people crossing the bridge at the time of the collapse, 145 are injured, and 13 are killed.

95

Middle of the night, I'm in bed, wide awake. This is what post-traumatic stress is to me. Once that gateway is opened, it stays open. In the safest place I can be, in bed, am I safe? Can I relax?

Deep . . . cleansing . . . calming . . . breath. Fill the lungs. Expand the chest. Slowly, let the air trickle out. Feel my back settle into the bed, feel my body let go.

Ask yourself: Can you do this? Can you let go?

Repeat the breath. Tell yourself: The citadel is safe, as safe as it'll ever be. Guarded by an invisible line of luck and hope and divine providence.

You reside in a sanctuary, an oasis, far from the hubbub and noise, a place of protection.

When insomnia strikes, I'll stop trying to drift off. I'll open my eyes and study the eggshell-colored ceiling, the straight square of it, the scalloped glass light fixture in the center like a buttoned button. I'll remember how I'd look at the ceiling after the collapse and wonder, How could that have happened?

I'd wonder lots of things.

How could it have fallen, really? That whole thing?

How was it I was on it?

How was it I am still here?

How was it I had no idea that big bridges fall sometimes?

I'll think about how much I hate that I know this. I'll think how much I hate that it's true.

Then, when I'll be too awake and my eyelids will lack the heaviness required to slide toward rest, I'll drift my palm toward my belly button. I'll feel the heat in my skin, the plane of my belly. I'll trace circles: first counterclockwise—small tight ellipses, then wider. I'll remember what my massage therapist and friend, Amy, taught me: the stomach is the vulnerable place, the emotional place, the place where people can pack away their pain, sorrow, longing. Then I'll be circling wide. Slow arcing shapes as if I'm tracing the edges of a mysterious lunar body. A circle that perhaps never ends. Infinity. With the alchemy of fingertips, I'll skim my rib cage. I'll comfort myself in this most simple and gentle way. And how I long to be comforted. The real me, who at night looks toward her ceiling and asks questions to nobody, senses that the whoosh-whoosh her movements make as her fingertips skim her belly is an echo of the questions she will always ask.

96

On August 10, 2007, the *Cleveland Plain Dealer* reports similarities between the collapse of the I-35W Bridge and the Ohio I-90 Bridge, in bridge design due to thinning gusset plates.

97

I needed a new exercise that wouldn't hurt my back. Working at the chiropractor and briefly with a personal trainer, I learned a handful of basic yoga poses with the goal of strengthening my back to supplement the PT I do myself. Those first attempts shocked me, how hard it was, just holding my body in a pose. I shook. My abs and arms burned. Legs ached. I'd become this inflexible? I couldn't imagine being more injured, like others. Just with my situation, I was overwhelmed. I continued with the yoga work three times a week and used the posture pump daily. Mornings especially, my cranky back ached. I hurt after yoga, but my mind benefited. I was glad I didn't have to return to hard impact running to get a workout.

I looked for a place to take classes and found Grasshopper Yoga Studio, located in a leased space of a church's second floor. On the website the instructors Silke and Nancy say: "As human beings, we have been given great capacities. Huge capacities are available for us to reason, to feel, to communicate, to visualize, to understand. Living in Yoga means truly realizing and fully opening to these capacities with grace, with calm, with joy, and with full presence in each blessed moment."

At the first class Silke said: "Notice what your bodies are doing. Feel the gravity. Now imagine your breath is a white light and you're breathing this white light down into your heart and then up into your brain." I imagined my breath and heart as bright, nearly nuclear white. And in that blinding light I visualized all the people who died. I felt panicky, holding back the urge to start blubbering. Keep it together. Instead, my eyes filled with pressure and salt. I bent my body, exhaled through a stretch, and looked at the smooth expanse of mint green wall. Trying to balance, I realized I couldn't think about anything else or I'd tip over. For six months I had walked around with ugly images and sounds in my head, reeling, unable to stop worrying, constantly vigilant and tense. I walked into that studio, and for ninety minutes I felt a peace I hadn't felt in months. Relief overwhelmed me with gratefulness equal to the heft of a life.

Yoga benefits people who've been through trauma. PTSD takes you out of your body, but yoga helps put you back in. Some poses instilled stillness as the only goal. Silke told us to stack three thick woolen blankets, lie back on them, and put the soles of our feet together. She demonstrated, splaying her knees outward. Then in that backward position, she told us, just breathe. I watched the ceiling fans, listened to the breathing of the other students in the room, and felt dampness on my brow. In that stillness I listened to cars outside on the street and the muffled piano practice in the next room, chords ringing through the wall.

There have been few places where I've felt safe since the accident—I mean, since the collapse. The word *accident* implies that the catastrophe was unforeseeable. Yet I work—always work—to balance my grown-up disappointment with my childlike needs, the simple requirement for hope. To believe in something good despite overwhelming evidence to the contrary. In my spare time I read a book called *Letting Go of the Person You Used to Be* by Yogi Lama Surya Das. He says: "Pain is inevitable, but suffering is optional. How much we suffer depends on us, our internal development, and our spiritual understanding and realization."

In yoga I'm learning lessons. One says: "I often know what my dharma, or 'path,' is and I do not do it. I often know what my dharma isn't and I do that anyway." Silke issued a challenge: try to find some stillness. Get quiet and balanced enough to hear or know your dharma, and do it. With feet spaced wide, I stretched taller to the ceiling, visualized myself strong and elongated like a weed, visualized my vertebrae spaced as they once were, restored to original health. Amid the quiet came ease and a growing awareness of my breath. I pondered the idea that my mind, body, and spirit were connected. When I got home and walked in the door, I continued until I reached Rachel's arms, tears loosening like change falling through pocket holes.

98

In attendance: all clients represented by the pro bono bridge collapse consortium and all consortium attorneys. On January 7, 2008, another meeting was held at the RKMC law office on the twenty-eighth floor. This was an important meeting to discuss the status of the investigation and proposed legislation and to answer any questions people may have. Wil's letter said,

"Given the Minnesota law governing attorney-client privilege, I'm sorry to say that friends and family (other than a spouse) may not attend."

Many legal meetings that I attended both with my individual lawyer as well as the pro bono consortium were termed "Attorney-Client Privileged Communication," meaning that the attendance and content of the meeting were only allowed by the client and their legal spouse. Although Rachel and I had legally married in Toronto, we were not recognized by Minnesota law or United States law as spouses.

I would have to go without my Rachel. But did I have to? I wrote to Chris Messerly.

FRIDAY, JANUARY 4, 2008

Dear Chris,

I received the call from Nancy [Remington, Wil's paralegal] asking that my wife, Rachel, not attend the lawyer meeting on Monday night.

I called Lisa Weyrauch [RKMC's paralegal] to clarify, "Are no husbands or wives attending?" She clarified that husbands and wives will attend but not boyfriends, girlfriends, or fiancés. Fine, however the difference between same-sex couples and opposite-sex couples is that for opposite-sex couples they may choose at any time to legally bond themselves to one another in MN by marriage. As a same-sex couple, we have no such option in this state.

As to attorney-client privilege, I don't really understand why there is so much secrecy, and I don't expect you to explain it all to me. Logically, I realize that you are interpreting the law. On a human level, I appreciate efforts that were made to "be sensitive," however until you or others are in my shoes, I don't think you can understand why it stings like it does.

Rachel is my spouse.

We would have a legal or civil union in MN if we could—we had a ceremony in MN (2006); we have a legal marriage in Canada (2005). We have arranged as much paperwork as we can to note our legal ties to one another, and we would gladly do more if the opportunity presents itself. We both have wills and power of attorney. Aren't those documents enough to note our "union" on paper?

I realize the answer is probably no, but if it is, are there other issues (like next of kin) that our wills and power of attorney cannot meet if my spouse can't even come to a minor lawyer meeting with me?

Thanks,

Kimberly

The law did not see Rachel as having any relation to me because we were denied the right of legal marriage. I learned that, no, Rachel couldn't attend. We were separated in the collapse and separated in this part of the aftermath. I attended attorney-client privileged meetings alone. Despite not liking it, we had no choice.

Prior to the meeting, I had written the copy for the survivor website, and together with another survivor, Mercedes Gorden, and with the approval of the group, we fine-tuned the site and found a company who donated hosting and design services. The site had recently gone live—a huge accomplishment—and it was announced at the meeting.

The *Star Tribune* ran an article called "Survivors Go Online to Make Case for Compensation." This would be to fight for the full bridge bill, not the interim relief from last Christmas. We were told to expect that the bridge bill would go to the Senate floor for a vote by mid-January.

In the mail I received a copy of notice that my lawyer sent to Attorney General Lori Swanson. The letter said it was a "formal Notice of Claim(s), pursuant to Minn. Stat. §3.736 subd. 5." Reading this, the notice struck me as incredibly official, and Wil was now the smartest person I knew. Thank god for these lawyers. How could I possibly have navigated all of this? Despite the past, I felt lucky again.

99

NTSB *Preliminary Transcript*

On January 15, 2008, the NTSB held a preliminary press conference, where to everyone's surprise, the board announced its findings thus far on the cause of the collapse. It blamed the collapse on under-designed gusset plates.

What did this mean? Did it mean that poor maintenance had nothing to with the collapse?

Broadcast nationally via live web stream and carried on local and national television, the press conference had not been anticipated to happen until late 2008. The NTSB cited this as "an interim report." Here were remarks by National Transportation Safety Board chairman Mark V. Rosenker.

Details of what the NTSB said are important, so here is the full text, but essentially this is how the fluke "design flaw" and collapse cause exonerating maintenance are described.

JANUARY 15, 2008

REMARKS BY NTSB CHAIRMAN MARK V. ROSENKER FOR I-35W BRIDGE PRESS CONFERENCE

Good afternoon ladies and gentlemen and thank you for coming.

As we all remember, on August 1, 2007, the city of Minneapolis suffered a tragedy when a bridge carrying Interstate 35W over the Mississippi River collapsed, killing 13 people and injuring more than 100 others.

The NTSB dispatched investigators that evening and we continue to investigate the accident. We released the accident site to the Minnesota Department of Transportation on October 12, and completed our on-scene investigation on November 12, although we have retained control of important portions of the bridge and might need to revisit the scene. Since the early days of the investigation, we have been working with the Federal Highway Administration and the Minnesota Department of Transportation to develop finite element analyses of the bridge to help identify the stresses in the bridge components resulting from the loads on the bridge at the time of the collapse.

I want to make it clear that we have not yet determined the probable cause of the accident, but there has been a development that I want to share with you today. We are issuing a safety recommendation to the Federal Highway Administration, which will be available at the conclusion of this conference and will be placed on the Board's website shortly.

First, a little history. This bridge was built during the mid-1960s and opened to traffic in 1967. The collapsed portion of the bridge was a steel deck truss, which was considered "fracture critical" because the load paths in the structure were non-redundant, meaning that a failure of any one of a number of structural elements in the bridge would cause a collapse of the entire bridge. There are approximately 465 steel deck truss bridges in the National Bridge Inventory, according to the Federal Highway Administration. In the years since it opened, the I-35W bridge experienced two major renovations, in 1977 and 1998. As part of these renovations, the average thickness of the concrete deck was increased from 6.5 inches to 8.5 inches, and the center median barrier and outside barrier walls were increased in size. These changes added significantly to the weight on the structure. On the day of the collapse, the bridge was undergoing repaving operations and there was construction equipment and material on the bridge.

The deck truss portion of the bridge consisted of steel beams that were connected to each other at 112 nodes, or joints, by gusset plates. There were two gusset plates at every node, and consequently there were 224 gusset plates on this bridge. During the wreckage recovery, we encountered fractured gusset plates from eight different nodes located in the main center span; all 16 gusset plates from those eight nodes were fractured. The other major gusset plates in the main trusses were generally intact. Gusset plates are generally designed to be stronger than the beams they connect and one would not expect to find them fractured. The damage patterns and fracture features uncovered in the investigation to date suggest that the collapse of the deck truss portion of the bridge was related to the fractured gusset plates, and in particular may have originated with the failure of the gusset plates at one of those eight nodes. As you might imagine, this created questions about the materials used to construct the bridge. However, materials testing performed to date has found no deficiencies in the quality of the steel or concrete used in the bridge.

Subsequently, the Safety Board and FHWA conducted a thorough review of the design of the bridge, with an emphasis on the design of the gusset plates. The investigation has determined that the design process led to a serious error in sizing some of the gusset plates in the main trusses; specifically, the gusset plates at the eight nodes I mentioned earlier. Basically, those 16 gusset plates were too thin to provide the margin of safety expected in a properly designed bridge such as this. These gusset plates were roughly half the thickness that would be required—half an inch thick rather than an inch thick.

In an effort to determine why those gusset plates are undersized, we wanted to examine the bridge's design methodology used in the 1960s to verify that it was sound. Unfortunately, the calculations used for the main truss gusset plates could not be found, so we cannot determine whether the error was a calculation error, a drafting error, or some other error in the design process.

We also examined the bridge inspection records and, although those inspections identified and tracked some areas of cracking and corrosion, at this point in the investigation, there is no indication that any of those areas played a significant role in the collapse of the bridge.

It is important to understand that bridge inspections would not have identified the error in the design of the gusset plates. The National Bridge Inspection Standards (NBIS) are aimed at detecting conditions such as cracks or corrosion that degrade the strength of the existing structure; they do not, and are not intended to, address errors in the original design.

During the course of our investigation, we have determined that errors of this nature have little chance of being discovered after construction. Routine design calculations that are performed for bridge modifications and special use permits do not typically include the level of detail that would be required to discover an original design error of this nature.

The Safety Board is concerned that, for at least this bridge, there was a breakdown in the design review procedures that allowed a serious design error to be incorporated into the construction of the I-35W bridge. The bridge was designed with gusset plates that were undersized, and the design firm did not detect the design error when the plans were created. Because of this design error, the riveted gusset plates became the weakest members of this fracture-critical bridge, whereas normally gusset plates are expected to be stronger than the beams they connect. Further, there are few, if any, recalculations after the design stage that would detect design errors in gusset plates. Finally, other programs to ensure the safety of our nation's bridges, such as the methods used in calculating load ratings and the inspections conducted through the NBIS program, are not designed or expected to uncover original mistakes in gusset plate designs or calculations.

It is important to note that the Safety Board has no evidence to suggest that the deficiencies in the various design review procedures associated with this bridge are widespread or even go beyond this particular bridge. In fact, this is the only bridge failure of this type of which the Safety Board is aware. However, because of this accident, the Safety Board cannot dismiss the possibility that other steel truss bridges with non-redundant load paths may have similar undetected design errors. Consequently, the Safety Board believes that bridge owners should ensure that the original design calculations for this type of bridge have been made correctly before any future major modifications or operational changes are contemplated.

I want to repeat that we do not yet know what caused the I-35W bridge to collapse that day, nor do we have indications that other similar bridges in this country have the kind of design flaw we found in this one, but we think that in the future, before major work is performed on a "fracture critical" bridge, all structural elements should undergo load capacity calculations.

I'll take some questions.

100

The consortium pro bono attorneys acted as watchdogs, advocates for the innocent. On January 16, 2008, the attorneys offered their perspective on the NTSB news conference. They told us: Please keep in mind that the state is part of the NTSB investigation. The NTSB and the state refuse to share any aspects of the bridge investigation. It is reasonable to assume that by blaming only the original designers, the NTSB knows that if the substance of its press conference were to be believed, it would serve to exonerate the state and leave the victims without any recourse as the statute of limitation expired long ago as to any claims against the original designers.

When Chairman Rosenker scratched his head, saying he'd never seen it before, this "design flaw," I believed he knew exactly what he was doing—suggesting that the blame could be placed on some nameless, faceless, out-of-business designers, Sverdrup & Parcel, on whom the statute of limitations had run out. A torrent of expert opinions about whether it was ethical for the NTSB to speculate, knowing it could have serious implications—especially for survivors—began to be voiced. Some faulted Rosenker for leaving an impression that nothing could've been done and that poor upkeep of the bridge was unrelated to the collapse.

Our situation reminded me of the Erin Brockovich story. That class action lawsuit depended on a smoking gun: did PG&E Corporate in San Francisco know about PG&E in Hinckley? Did they turn a blind eye to the evidence? It was a case of corporate greed, profits, and cutting corners. In the movie Erin met a former PG&E worker who had been asked to shred documents. A chilling moment occurs when the worker tells Erin, "I wasn't a very good employee." He had saved a 1966 letter that proved PG&E Corporate in San Francisco knew the water was being contaminated at PG&E plants in Hinckley, California, which resulted in an abnormally high incidence of brain tumors and disease. In 1996 the lawsuit was settled for $333 million—the largest settlement ever paid in a direct action lawsuit in U.S. history.

A reader comment from an article in the *Cleveland Plain Dealer* newspaper summarized the unofficial feeling: "Typical of the pathetic commitment to transportation infrastructure in the United states in recent times . . . be it maintaining roads & bridges or mass transit and rail. Perhaps things would be in better shape if we weren't flushing untold billions down the toilet in Iraq."

While this was one sentiment at the time, did it have any merit? Thinking critically, we realize that people who are Republicans, the anti-tax party, had to drive these roads too. Why would they choose to leave bridges in disrepair?

Because raising the necessary funds via taxes is unpopular and doesn't get people elected to office.

Because we've fallen so far behind that catching up feels insurmountable.

Or maybe because it is less important what's true and more important what you can get people to believe.

After the NTSB preliminary announcement, I, along with another survivor, interviewed again with the media—this time with the *Star Tribune*. I was quoted saying: "The fund isn't dependent on blame. It's about doing the right thing." The fund would be an alternative to lengthy and expensive court battles. After the NTSB made its announcement, I learned about the 1993 inspection report (mentioned earlier) that documented MN/DOT's knowledge. To review: that 1993 report, in all its eerie specificity, revealed that the gusset plates were thin, with a "loss of section 18" long and up to 3⁄16" deep (ORIGINAL THICKNESS = 1⁄2")."

Part of me felt like, *Ha! Gotcha. Liars.* But another part of me just felt devastated. The NTSB: it said it hadn't known. The board, specifically Mark Rosenker, said it had never seen it before. But it had. From Ohio's I-90 to MN/DOT's own inspection reports, the NTSB had seen it before. But who else would know this? If your industry is not engineering or transportation, guess what, it's not common knowledge.

The NTSB wasn't supposed to announce its findings until they were final. Announcing the findings early had strategic merit: give the public time to get used to the idea of a "design error." *Give them six months to think about it.*

The NTSB. Was this powerful agency influenced by the governmental powers in charge? I ask myself because why would they stand in front of (essentially) the world and essentially tell us that this bridge fell because of *a fluke* that *nobody* could have predicted?

Post-conference, there were few other official voices who could outspeak the NTSB. Legal action was pending with the state, to be followed by the potential of further legal action with third-party entities, which could take years. In the meantime the NTSB had spoken; seemingly, its indication of cause would rest, as more and more people would get used to the idea. This is the NTSB, after all, whose investigators use science and unbiased facts to get to the truth of the causes of all kinds of disasters.

The truth is, I would eventually understand that the bridge collapse wasn't caused by a single factor but a *combination* of factors. But it turns out it wasn't that simple. It was more complicated, and it would be nearly a decade before I could be healed enough to face the reality that they had known. The NTSB, and MN/DOT before them, had known.

So. Let's review. Here's what I know so far. The NTSB says the collapse cause isn't maintenance; it's this gusset plate the inspectors never saw. Based on my research, at this point there was deferred maintenance that might not have been sufficient to cause the bridge to collapse all by itself, but there's a record of deferred maintenance, and the inspectors knew about the size of the gusset plate, so I'm not satisfied with this answer, and I'm going to keep looking. I wouldn't know this then, but later—*holy crap, I couldn't believe it*—there will be more. The full collapse truth is yet to come.

101

"Bridge Failures in Ohio, Minn. Linked": January 2008

WASHINGTON, Jan. 19 (UPI)—Sagging on an interstate highway bridge in Ohio in 1996 was caused by problems like those investigators found on the collapsed I-35W bridge in Minneapolis.

Preliminary findings released this week by the National Transportation Safety Board said sixteen of the Minneapolis bridge's gusset plates fractured. The plates connect beams on the bridge.

While a Federal Highway Administration report that was also released at an NTSB news conference mentioned the I-90 bridge in Ohio, agency officials suggested the problems with the Minneapolis bridge had never been seen before, the *St. Paul Pioneer Press* reported.

The NTSB did not investigate the Ohio bridge closing. An Ohio Department of Transportation report blamed the sagging on beams that "were not adequate to support the design loads of the structure."

While the NTSB is known for its investigations of plane crashes, it has less experience with structural failures, the newspaper said.

Thirteen people were killed and scores injured when the I-35W bridge collapsed suddenly at rush hour Aug. 1, 2007. The sagging of the bridge over the Grand River in Ohio caused no injuries but the closing caused massive inconvenience.

102

"NTSB Chief Softens Comments about Bridge Collapse": January 2008

WASHINGTON, Jan. 29 (*Minneapolis/St. Paul Star Tribune*)—The chairman of the National Transportation Safety Board backed off Monday from earlier remarks that seemed to rule out all but design flaws as the "critical factor" in the Interstate 35W bridge collapse.

Mark Rosenker's clarification, made to U.S. Rep. Jim Oberstar, Minn., came as NTSB investigators again contacted one of the researchers involved in a study of an Ohio bridge that buckled a decade ago due to undersized gusset plates similar to those that have become the focus of the 35W investigation.

Rosenker, meeting privately with Oberstar, said he did not mean to suggest that the finding of undersized gusset plates on the 35W bridge reflects the board's final conclusion on the cause of the Aug. 1 accident. That conclusion is expected later in the year.

Their meeting, and a separate letter to Oberstar, came a week after Oberstar criticized Rosenker for rushing to judgment on the cause of the collapse, calling his comments two weeks ago "highly inappropriate."

Rosenker, briefing reporters on Jan. 15, cited 16 undersized gusset plates that joined the 35W bridge's steel beams together as "the critical factor" in the collapse. He also said gusset plate design "tells us why the bridge collapsed."

103

Before the House can pass its version of the bill, it had to pass many committees. If one committee killed the bill, it was over. So I spent my free time composing an opinion essay in response, making sure the message was on point, and trying to find a home for it. The *Twin Cities Daily Planet* published my essay on its website, in which I urged: "Can't we do things differently? Let's look to the future and give our bridges top billing. It's better than the alternative." I noted the survivor website www.35wbridge.com, where anybody who wanted to help could access links to correspond with their legislators.

The consortium hosted a Bridge Victims' Day at the Capitol on February 25 to lobby for a survivors fund. We met at 11:00 a.m. in Room 300 North—State Office Building. Survivors, lawyers, and the sixty-one kids from the Waite House who had been on the school bus were there, and everyone brought friends, family, and anyone else who supported our efforts to establish a fund. As a group, we wore bright red for unity and to stand out. A half-page flyer was circulated and distributed, titled "Bridge Victims' Fund Now." It showed a picture of the wreckage with all the destroyed concrete and a single woman sitting upon it, a woman I later learned was Thuy Vo, who, despite not possessing the ability to swim, had fought and made her way to the broken concrete. Centered at the bottom: "For more information visit www.35wbridge.com." There were three simple points:

This is a UNIQUE man made disaster of catastrophic proportions.
13 dead. More than 100 injured. Tens of millions of dollars in wage loss, medical expenses and destruction of lives.
Support bill without individual caps on damages. (Support H.F. 2553)

With all these hurting people gathered in a room, the intensity was palpable. The lawyers said it was possible that our politicians might address our group. But before I go further, I have to briefly tell you about one of the thirteen victims, Peter Hausmann. Peter's body was found partially inside the car of Sadiya and Hana Sahal, and it looked as if he were attempting to save them. But what I didn't initially realize was that at least one witness reported seeing him walking around on the bridge after it collapsed! To think... he would have probably lived and been a "survivor" like the rest of us. I couldn't imagine what he went through, and his noble last act amazed me. To think, his last moments on earth were spent trying to pull another from the murky water.

That day I sat by Peter's widow, Helen Hausmann. At one point Helen said, "I'm just so angry," and she lowered her face into her lap and cried. I had just met her, but I started crying too, and for a beautiful horrible moment I reached out and instinctively rubbed her back. Usually, I would've been self-conscious in public touching someone I had just met, but it was one person reaching out to another. I couldn't do a thing to help her, just witness with her. At a later meeting Helen searched me out to tell me she was going to see her relatives in Kenya for four months. It was the first time I had seen her smile.

The Speaker of the Minnesota House of Representatives, Margaret Anderson Kelliher, made a point to address us, showing her support by her presence and by speaking. She was human. She cried. She told us how sorry she was that this had happened to us. I found it incredibly moving, and judging from the emotion in the room, others did also. She expressed her support for our efforts and promised the legislators would work to make sure a collapse never happened again. It was reassuring to hear from someone in charge, saying this clearly instead of all of the excuses we'd been hearing from MN/DOT and others. She was the first official to apologize for this happening, which struck me as remarkable and brave, and it was just the message that people yearned to hear. Empathy, conscience, compassion.

104

Yoga class started at 8:30 a.m. I exited the freeway at Thirty-First Street and Lake and waited at the red. Running late, I tapped my fingers impatiently on the steering wheel. Come on come on.

A man stood at the curb holding a tattered cardboard sign. "Homeless. Really bad. Help me." Sometimes I don't have cash. Other times I debate: if I give cash, does it mean I'm generous and kind or a sucker? That morning I looked at the man's long face and wondered why I think twice. Inside the middle console, I fumbled for coins and found a five-dollar bill. Out of the corner of my eye, I saw the man approach the car. A stranger. I took a deep breath, unfolded the bill, and held it out the window.

"Bless you. Thank you," the man said, nodding and bowing. I nodded back, hoping I looked encouraging. Without words, with tears in my eyes, I drove away, figuring I'd soon forget all about him.

At class I arrived a few minutes late, flipped off my shoes in the space outside the door, opened the door gingerly, and tried to minimize my disruptive entrance. Silke was talking to the class about nostril breath. Around ten students sat on their mats, hands at their noses, alternately plugging a single nostril at a time. Depending on what side of your nose feels most clear, Silke said, it could mean certain things. If the pathway of your breath was most clear on the right side, it could mean *left* brain stimulated, high blood sugar, energy, creativity. On the left, *right* brain stimulated, quieter, low blood sugar, passive. Ideally, if both sides were clear, you had a more balanced state.

My nostril felt clearest on the right.

After warming up, we started asana, or posture practice. We cycled through poses, starting with child's pose: where you kneel, fold your chest over your thighs, and rest your forearms and forehead on your mat. My thighs and hips felt stiff. Silke said, "Breathe in and come to all fours." The students rounded their backs and dropped their tailbones and heads. "Breathe," she repeated. Then we arched our backs and looked toward the ceiling. We returned to child's pose, then flipped our toes under and lifted into downward dog. Twisting poses like Warrior II—a deep lunge in which we reached with one arm over our thigh—were harder for me. Once in the pose, Silke told us to twist more, chest up and back, if we could. I tried, but my torso stayed rigid like a brick. I wobbled. The disc in my lower back pulled with a ripping sensation.

Holding the pose, my mind drifted. I returned to downtown Minneapolis, the corner of Sixth Street and Hennepin Avenue, people-watching from a bar stool. Commuters, youngsters, businesspeople dressed in suits—people in varied shapes, sizes, and ages—crossed the street, looking content. Did they know that day would be but a memory soon? That present moment would fade like newsprint? I might've died twice now. My life, your life, we were at the mercy of fate, luck, or a greater power. I've often wished to stop time. Let me freeze it. Let me catch my breath. Let me be safe and sound for such a long time that I get bored with it.

I had briefly looked away from the street. When I looked back, a pack of cyclists had taken over. The normal cars and trucks were replaced with hundreds of bikes. At the red light the cyclists waited, dressed in tight aerodynamic suits and tear-shaped helmets. When the light changed to green, one by one girls and boys pedaled away. Symphony of bones and blood, sinew and muscle, hands and feet worked in chorus, pushing through the intersection. I watched them pass. Ephemeral flash. Then gone. So many tires, so many faces, so many circles.

Next we moved into a longer sequence beginning with a sun salutation. We stood toward the top of our mats. Starting with palms together, we then swooped our arms outward in an arc. Arms stretched above our heads, we visualized the sun. Silke said, "Visualize the light outside of us, the light inside of us." Behind my eyes, burning. I imagined the light. At first warm: orange and yellow. Then brighter like the light described when people pass on, only this light burst with vitality. Like I often did, I remembered those who had died on the bridge. Silke continued: "Fold into a deep forward bend." I did my best, my back like a board, remembering when it used to fold this way or that, supple as a blade of grass. From there we did plank pose, cobra pose, downward-facing dog, hopped our feet between our hands, and finally, with a sun salutation, we began again.

Before the end of class, we transitioned to Savasana. *Sava* in Sanskrit means "corpse," and *asana* means "pose." Put them together, and you had corpse pose. In my first class we reclined on our mats, legs extended, arms resting, eyes closed, and I thought, This is a pose? It's like I'm sleeping.

Silke told us, "Feel the weight of your bodies, the ease and the stability of the earth beneath you."

L5 ached. I bent my legs and pointed my knees to the ceiling, relieving the pressure. I closed my eyes, breathing, resting, when Silke appeared with a mountain of blankets.

She whispered, "I'll put these under your knees."

"Thanks," I whispered. *How thoughtful*, I said to myself, as she adjusted another folded blanket under my feet. Then I relaxed all of my body: legs heavy, mind calm. All of us in Savasana, abandoning care to the wood floor. Eyes closed, I listened to the quiet in the studio, to students making slight adjustments with their bodies. My mind stilled, and I realized I was fully present—an accomplishment. I felt peaceful and, for a moment, safe.

I thought about the stranger on the street this morning, remembered the look on his face when I gave him just five dollars. I thought about Silke with the blankets, how it eased the pain. And *flash!* I realized it made a circle. My small kindness had come back to me.

105

My height. I was re-measured on April 3, 2008. No change. Since the collapse, our dog, Lucy, has been my next-best friend behind Rachel. The poet Stephen Dunn said, "A good dog in your house can make you more thoughtful, even more moral."

We adopted Lucy on a snowy February morning in 2004, the day after Valentine's. She'd had two homes before us, and her most recent family had lost their home in a fire. Adopting her was easy. Keeping her was another story. At first Lucy had severe separation anxiety, which made housebreaking tough. For a while I tried to remain unattached, reminding myself she might not work out. Shortly after her adoption, I drove down a six-lane freeway with her, and a noxious plume of smoke wafted across the lanes. Where it had come from I wasn't sure, but the burning smell intensified. In the back Lucy panted and paced. Why wouldn't she calm down? I continually peeked at her in my rearview mirror when I saw liquid running from side to side across the floor. Pee, I realized. But at first it didn't click. Was something wrong with her? Then it dawned on me: she remembered.

She and I: survivors. What had she seen? What had she gone through?

Once, on the Fourth of July, I told Lucy, "C'mon, let's go potty." She rose from her dog bed and plodded forward, as if each paw were hardened in cement. These words had worked famously once, but now they did nothing. As we walked to the back door, I gained on her. I glanced back; she was stopped by the piano.

"Come on, honey. Let's go," I coaxed, using words for beloved sweethearts or children. "Come on, pumpkin."

I tried a different tactic. With authority I said: "C'mon. Let's go. Now."

She froze like winter. I had to trick her. I grabbed her leash, tags jingling. Surely, she'd come for a *w-a-l-k*. But no. She submitted on her side like a scoop of mashed potatoes. I lifted her sack of Idaho spuds. Finally, she was up and moving. I heard the neighbor's dog barking outside and said: "Go see Sparky! Go potty!" I opened the door, led her out, and waited for her return. One, two ... too quick, already a scratch at the door, so I let her in. There had been no fireworks but in memory.

Most nights, after work, dinner, and cleanup, I grabbed poop bags, cookies (Milk-Bones), and Lucy's leash, and together we meandered along tree-lined sidewalks, routes we'd traversed hundreds of times. On one walk we turned toward home and saw a neighbor burning brush in a fire pit. Smoke wafted across the lawns, and before I could wonder, Does she remember? Lucy glanced up at me with golden worried eyes. She held eye contact longer than usual as if to ask, "Am I safe?" I nodded. Yeah, you're safe, I said without speaking, compassion extending from human to animal. She stared back at her paws, and I registered a wave of recognition. She and I: both of us, with post-traumatic stress. Lucy's visceral memory showed in her eyes and her plodding, the way she pushed forward despite smoke and bad memories.

106

A flurry of hearings occurred from February to May, often several per week. As I became more obsessed with lobbying, I felt a shift. I went toward the families of the thirteen—and I looked forward to progress. When it eluded us, it gnawed at me. I relished any evidence that I was appreciated. For some reason I needed it. (It wouldn't be until much later that I'd finally recognize this need and name it: *survivor's guilt*.) I was only one of many players in the background, working. Each day I longed for some kind of milestone, some way of speaking to make things better. But I often forgot that I was a survivor too. In some ways it separated me from the others and isolated me. Survivors didn't understand it was a way of coping. Taking action felt better than getting socked.

At the Capitol I met Mohamed Mohamed, a relative of twenty-three-year-old Sadiya Sahal, one of the thirteen people killed. When Sadiya died, she was five-months pregnant. In the car her daughter, Hana, two years old, and an unborn seed that never ripened. Why wasn't her unborn baby considered the fourteenth victim? Apparently, thirteen was the number that would stick. After we talked in a marble hallway, a smaller offshoot to the grounds, ambient light shone against the walls and pulsated in the sun. We spoke infrequently. Surely I asked, How are you holding up? I'm so sorry. That kind of thing. He said he was doing okay, it is difficult, but Allah deemed this to be so and he must accept it.

Mohamed Mohamed, man of two same names. I imagined sitting longer with him, not knowing what to say. While we know that words can't undo, they could fill space. I would want some words. And they'd be the trivial kind. The weather. Bemoaning it too hot, too cold, or how perfect the weather was if only it could stay this way. And in these trivial words they'd overarch the theme of loss: if only. If only Sadiya and those children had been five minutes late. If only Allah had picked someone else—or no one?

When I made it across the busy avenue and arrived at the meter, I viewed Mohamed in my mind. I placed a hand on his and held it there for an extended pause. It wasn't nearly enough, I knew—and touching him might be offensive in Muslim society because I'm a woman, so I knew it was merely a wish. But this touch, in our hands-off society, would convey my grief.

Sadiya. Such a pretty name.

107

Both legislative bodies voted on their versions of the bill. When the Senate voted, survivors had spent a full day attending hearings. There was a delay before the Senate vote, and we were told it would be several hours. We scattered to pass the time. Hours later I got a voicemail—the Senate had reconvened. I rushed back to the Capitol and worked up a cold sweat running to the Senate chambers. When I arrived, survivors were exiting the upper-level gallery that overlooked the floor, where the legislators did something unprecedented: they gave a standing ovation and applauded us as a show of support. I had missed the vote.

I walked back to my car, and once alone, I started sobbing. I had persisted in my commitment and buried myself in the issues, focusing on fixing what was wrong externally. I went to plenty of doctor appointments addressing my own situation, but with the bridge work to do—from letter writing to attendance at survivor meetings to just generally obsessing about ways to make a difference and improve a bad situation—I had put so much of myself into this. I was crushed to have missed it.

When the Senate and House votes were finally achieved, it seemed like our work was done and we could move forward. But the House and Senate versions didn't match. We learned that the next step was for each bill to go to conference committee, where joint members from both legislative bodies would attempt to reach a compromise. The two versions had to be merged to create a single, final version. The first meeting of this conference committee was March 26. Meetings continued into April.

I wanted to be involved, to help. But weeks passed during which there was nothing that anyone could do. No letters to write, no emails to send, no hearings to attend, no lawyer meetings to digest. The legislators debated behind closed doors, and we learned of progress through our lawyers and the media. Finally, Chris Messerly said I could help by writing handwritten letters to the conference committee members. The letter would accompany a video from a news story that ran on a local TV station about the Coulters, a family of four, all survivors, who had sustained severe injuries. To mention only a few: broken backs, brain damage, and for Paula Coulter months of recovery at Courage Center. We were fighting for a final bill without caps.

I wrote: "I survived the bridge collapse, and I don't support individual caps. By great majority others don't either. I'll never know why I was spared the catastrophic injuries others suffered. And I still don't understand how a bridge can collapse here, at home. What is clear though is this: you have an important decision to make. Please watch this video. I hope you conclude that the fund can be flexible in the way it's distributed. You can choose to make all victims as whole as possible." I signed it in the way to which we'd become accustomed: "Thank you, Kimberly J. Brown, survivor, in car B3."

For months it seemed the bridge bill went nowhere. The lawyers did their best to keep us informed of the convoluted legal processes. The politicians weren't agreeing, and the bill seemed to be in real danger of never becoming law. All our hard work, all the people who'd organized and fought for us—the lawyers, politicians, community members, and each other. Would the bill reach fruition? Would the outcome lead to long-term healing, an investment in the future for not just the concrete and steel but also the people?

108

Spring popped the flowers open: at every turn, on every block—and in the gardens near the boulevards. Lucy trotted to the end of the retractable leash, and I followed to the crunch of her padding over rustling leaves. We passed several lilac bushes when the urge overcame me to stop at one. She took a breather, and near lavender clumps, I leaned in and breathed deeply.

At night she curled into her dog bed and turned into a cinnamon roll. I slipped naked between cool sheets, and we left this place and went to dreams. Memory transported us back to explosions and bright lights. Her haunches twitched like lightning contacting ground, her paws like chocolate chunk cookies skittering from the unseen. Eyes three-fourths open, her rib cage rose and fell between the shocks, on time like claps of thunder. Sometimes we looked for rescue, and sometimes it came. Both of us with our PTSD, unlikely partners.

109

October 2007 had marked the start of many months pouring my energy into advocacy. On May 8, 2008, the governor of Minnesota, Tim Pawlenty, signed a $38 million bridge bill into law, expediting compensation to survivors and victims who chose settlement and waived their right to sue the state. This part of the struggle was finally over. I started carrying the memory of the thirteen who died as if they had been my brothers, my sisters, my parents.

When I've met bridge friends who live now without a family member, I feel inexplicable guilt. But it's more than that: a cosmic knowledge that we get one shot. Use it wisely or waste it? I no longer accept anything as being written in stone. Nothing is so permanent that it can't be analyzed, recalculated, judged in a different light. Obstacles in life become opportunities: to notice, to slow down or speed up, to love or long for it. It's a blessing to be here to feel and to learn and to not be doomed to repeat past mistakes and live with regret. Those thirteen killed didn't get that chance. I'll live my life remembering them. You don't have to know them to do the same.

The moment the governor signed the bill marked the end of lobbying. I had missed the standing ovation and passage, but I was present for the signing. If you had been with me, if you had been me, you'd know that I felt like a runner, like I'd gone on a long-distance jaunt, my shoes worn down, my brow beading with sweat, and there I was, with exactly enough breath to cross the finish. Although I was there, in the present, I could already suspect how the days would feel in the future, how they would disintegrate behind me, like crumbs, broken apart and forgotten. But right then, after the ink had dried and our footsteps upon the Capitol marble had faded, the true work for me began. How would I reclaim my life? Advocacy forced me to postpone some of my own work. Regular chiropractic and massage appointments were doable. But addressing the emotional and physiological effects was more difficult. How would I face my own mortality, renew my hope, live with purpose despite fear, get over the anger? Was it even possible to forgive? What now?

110

I've always liked documentaries. As a kid, Mom and I would settle in atop the floral bedspread with our cats and maybe a blanket and watch television shows about mysterious lost civilizations: ancient Egypt, Pompeii, King Tut. So it was a natural inclination I followed when, post-collapse, I began watching documentaries about topics that were mysteries to me: engineering and specifically bridges. *Nova: Trapped in an Elevator* gave me facts to assuage an intellectual fear of falling. *Modern Marvels* on the History Channel guided me through the elementary principles of corrosion in metals. And particularly useful in understanding the science of bridges,

The Crumbling of America, also on the History Channel, turned out to be an amazing resource. James Meigs, executive editor of *Popular Mechanics*, said, "Mesopotamia, The Incas, The Indus Valley, all those societies started with dikes and levees and agriculture and roads . . . and without that, humans would never have risen above the hunter gatherer status."

111

Back at yoga. Eyes closed, legs crossed, palms together—during the usual end-of-class meditation, Nancy said, "Always be willing to start over." What? My ears perked up.

Always be willing to start over.

I hadn't considered my own mortality in a meaningful way. As a kid, I had focused all my energy on Mom, on managing setbacks. At times it upsets me to be so hyper-aware, and I feel small, and even doomed, knowing I can never go back. Other times I feel like I have a leg up, an unfair advantage. I imagine whispering in a stranger's ear: "Hey, wake up. It doesn't last forever." I imagine all that we could do with that power, that knowing. Travel; view life differently; be creative; never be trapped at a job you hate; work your hardest to create the life you want; be kind, generous, thoughtful; surround yourself with worthy people; save lives, be it people or animals; tell people you love them, see them; take lots of photos; cook good food; eat good food; wear clothes that feel good; enjoy days of mild weather.

It struck me: this is what this bridge has done. I would look forward. Lobbying had ended. It was time to start over.

Those seven months of hard work exacted a toll. After the governor signed the bridge bill, I stopped. I shut down. I realized it's over. It's finally over. All these people would receive all this money to help them recover because my letter started it. Don't get me wrong, my involvement was a fraction of the big picture: many people did copious amounts of work. But because I spoke, in a poetic reversal of my childhood silence, big things happened.

I don't know where we get the idea, but we get it somewhere. The idea that somehow we will live a comfortable life. Grow old. Follow a specific

path, then die a merciful death made comfortable with drugs in old age. That's not what's happened for thirteen of our friends and neighbors.
   Start over.

112

The narrator in *The Crumbling of America* said: "Bad things happen when an empire allows its infrastructure to decline."

113

I designed a prototype of a bumper sticker with all thirteen names.
   I was trying to manage notifying the families regarding the bumper sticker when I got word that one of the family members "[wasn't] too comfortable" with it. I emailed that family member separately to express my support and to ask if she would share her concerns. I was feeling a tad discouraged since

12. Abandoned bumper sticker. © 2008 Kimberly J. Brown.

I had felt so heartened that this would be a healing thing, but it was amazing how complicated (what seemed like) a simple thing could be. I also hadn't received responses from two-thirds of the people. So, while I wanted to give it more time, I was concerned. If I didn't or couldn't get approval from all families, I was trying to find out if there was any legal implication to going ahead with the sticker (i.e., would I be breaking any laws?). But more than any of that, my gut said that if I couldn't get approval from all families, I couldn't feel completely confident about making the stickers for everyone.

One of the victim's widows said that I probably wouldn't receive 100 percent agreement on something like this because of all the emotion involved. He personally didn't have any problem with it because it was a way to remember those who perished on August 1. He suggested perhaps I pass it by the pro bono lawyers to ensure there were no legal problems and after that, if everything was okay, I could make it available to all and ask for donations to offset the costs of printing or ask the printer to donate them in support of those most harmed by this disaster.

I didn't know what it was like for family members to talk to someone who had lived, but I expressed to the families that I just wanted them to know that whether this sticker got made (or not), I cared, and I remembered them.

So, after checking with Chris Messerly (one of the pro bono consortium lawyers) and Wil (my lawyer) and after putting a call into a lawyer who works with privacy law and after having done some of my own research, I concluded that the sticker with the victims' names wasn't going to work. This made me melancholy, since I felt the power came in remembering each of them *by name*. I also felt that the sticker had meaning because it "spoke" with many messages. The bridge shouldn't have fallen. These people should be remembered. The bumper sticker said where and when this happened and marked that time in history. It said we should honor them by cherishing our lives and by trying to make something of it, to not take it for granted.

But I didn't want to hurt anybody. In the end this version of the sticker with the names never came to fruition. In an effort to inform every family, I was unsuccessful in gaining the consent of all families. Knowing that they were grieving, I didn't push. I keep asking myself, Why did I care?

Why put myself through this? I just had to. Not much of an explanation. I had broken out in hives again. I had had them for weeks after the collapse. I guess I was really trying to make some kind of sense out of something that made no sense.

114

Dan McNichol, author of *The Roads That Built America*, said in *The Crumbling of America*: "I went to China, interviewed engineers—they couldn't believe that a bridge in America had collapsed. In China bridges collapse but not in America."

115

On June 23, 2008, I composed an email and sent it to the victims' families.

> I wanted to let you know that I've decided to cancel the making of the bumper sticker with the names.
>
> I apologize to those of you who were looking forward to having it. While I have five families who have approved, I've tried hard to contact all of the families and I haven't been successful.
>
> I wouldn't feel good about this without everyone's blessing, so I hope you guys understand. I hope all of you know (and feel) that I was trying to do a good thing.
>
> I might still like to try to make a modified version (without the names), but I'm not sure about the wording. The intent would be the same—to honor them. If you have any suggestions, I'd love it if you'd send them my way.
>
> Again, you have my heartfelt condolences, and I wish all of you the best.

Bob Ross and I talked via email.

YOU ARE ALWAYS SO THOUGHTFUL AND WRITE SO NICE. I WISH I COULD EXPRESS MYSELF AS WELL AS YOU. I THINK THE ANNIVERSARY IS GOING TO BE ROUGH FOR EVERYONE.

I feel sad that this didn't work out. I had no idea what I was getting myself into. It started so innocently.

As to the anniversary, I think you're right. I feel pretty emotional these days . . . I think I must (deep down) feel pretty guilty about having lived, when so many others weren't so lucky or were horribly injured. I have had an outbreak of hives for the last three months, which my massage therapist and lawyer confirm is probably psychosomatic. Great.

I don't know what to do. Fact is, I can't do a d*mn thing to fix it. I tried to make a difference with my letter and getting the legislature to act. And I'm proud of my role in that. But I just feel so helpless. I think I keep trying to do things to make a difference to prove that I deserved to live. That's what I blurted out in therapy and it left me speechless.

At the same time, we all have to heal, somehow. We all have to find hope in the future, even amongst so much sadness and loss. There has to be a future for each and every one of us.

I'm supposed to talk with WCCO tomorrow afternoon. I have no idea what to say. I kind of wish I hadn't agreed. My main point will be to just say "thank you" on camera to everyone for the support and to those who donated to MN Helps.

I THINK YOU WILL FIND THE RIGHT WORDS TO SAY, AND IF YOU DON'T KNOW THANKING EVERYONE IS PERFECT. THERE WERE SO MANY NICE PEOPLE WHO WAITED, PRAYED, HELPED, GAVE THEIR TIME, HELPED WITH MONEY AND ON AND ON.

DON'T FEEL BAD FOR A MINUTE THAT YOU LIVED AND OTHERS DIDN'T. THAT WASN'T YOUR CHOICE. IT JUST HAPPENED. IT SHOULDN'T HAVE EVER HAPPENED BUT IT DID . . .

I SAW KIM DAHL ON TV AND WATCHED HER. SHE SAID SHE HAD JUST WAVED TO THE TRUCK DRIVER ALONGSIDE OF HER AND HE WAS HONKING AT THE KIDS, AND THEY WERE WAVING BACK, AND BAM IT WAS OVER JUST THAT QUICK. SOMETHINGS I THINK YOU CAN'T FIND MEANING TO.

BUT I ALWAYS HAVE APPRECIATED BEING INFORMED BY YOU, AND I AM SURE ALL THE OTHERS DO ALSO. YOU LET US KNOW WHEN THE MEETINGS WERE AND WHERE THEY WERE. THAT WAS BIG. SO DON'T EVER THINK YOU DIDN'T DO ANYTHING. YOU DID A LOT. I WAS THINKING OF YOU THE OTHER DAY WHEN I WAS EATING CHESTER CHEETO'S IN THE CAR. ME AND MY DOG. SHE LOVES THEM. SHE EVEN HELPS GET THE ORANGE COLOR OFF MY FINGERS.

TAKE CARE, STAY SMILING, BOB

Bob,

Your words mean a lot to me; brought tears to my eyes. Thank you. Take care, too.

P.S. Glad to see the all caps are back. It's not right when you write all little.

YOU'RE SO FUNNY. BETSY ACCUSED ME OF YELLING IN HER EAR WHEN SHE READS MY ALL CAPS SO FOR A LITTLE WHILE I WHISPERED TO PEOPLE, BUT PEOPLE WERE REQUESTING THE ALL CAPS.

Throughout the day I kept rereading this part of Bob's email: "DON'T FEEL BAD FOR A MINUTE THAT YOU LIVED AND OTHERS DIDN'T. THAT WASN'T YOUR CHOICE. IT JUST HAPPENED. IT SHOULDN'T HAVE EVER HAPPENED BUT IT DID." With my fist at my mouth and barely able to see through tears, I read it again and again and again.

116

Stephen Flynn, author of *The Edge of Disaster*, said in *The Crumbling of America*: "When we look at Rome as a great civilization, a big part of what made it so great was the infrastructure. The road systems, the bridges, the aqueducts. As the Romans started to neglect that investment, they crumbled and collapsed and we ended up with the dark ages."

117

I redesigned the sticker in the same colors and overall design, but I removed the names and instead used the text REMEMBER THE 13. A local company, John Roberts Printing, generously donated the printing fees. I paid for materials and distributed stickers to all of the families of those lost. I wrote a note for the back of the bumper sticker—a labor of love for many months and something I did for my own healing.

ABOUT THIS STICKER

I was on the 35W Bridge when it collapsed, and I designed this bumper sticker to honor the 13 victims who died. The idea for this started in March when I traded my old car for a hybrid. I wanted to put a 35W sticker on the new car, but I wanted to do something special.

That day I was on the bridge, and I felt it shake, the concrete rippling in front of me—I thought, "This is it. I'm going to die." For some reason I lived. I'll never know why. I have asked myself, "If it had been me, what would I have wanted?" I immediately knew the answer. "I wouldn't want to be forgotten."

This sticker is about them. I want them with me every day, at my back, in my mind, in my heart, watching over me. May "the 13" inspire and remind us of what life can and should mean, and may they be remembered.

I didn't know them, but I never want to forget them.

Kimberly Brown

Survivor in Car B3

June 26, 2008

From a survivor's spouse: "I do not know what I will do with all of them, but I think they are great and don't want to run out if somebody ever wants one. So I would like 25 of them if that's ok? If you don't have enough for everybody, I will take as many as I can get up to 25. But please make sure everyone gets theirs before I hog them. If nobody told you yet today, YOU ROCK."

From a survivor: "Hi Kimberly, I will take 50 stickers if they are available.

13. REMEMBER THE 13 bumper sticker. © 2008 Kimberly J. Brown.

We would all display them proudly in honor of our brother Jolly and the other 12 people who were taken that day."

From Pollee Chit (Vera's daughter and Richard's sister): "What beautiful words you used to describe your experience and your feelings about our lost loved ones! I cried when I read your email. My mom and brother would have been so honored to know you put so much thought into remembering them. I admire your creativity and thoughtfulness. I feel in my heart that they are our angels watching over us all. You are always in my prayers."

118

> No man ever wetted clay and then left it,
> as if there would be bricks
> by chance and fortune.
>
> *Plutarch (ca. AD 46–120)*

119

I searched for another EMDR-certified therapist, and during summer of 2008 I found Barb Stamp. Barb uses an electronic board with a science fiction name: the Advanced LapScan 4000—as the manufacturer describes it—"visual, auditory and tactile modes are synchronized." It's rectangular and about the size of an opened magazine. A series of green digital lights illuminate and move in various patterns. When Barb presses the power button, the lights move in a lemniscate—a sideways figure eight pattern. The pattern could be changed (e.g., to a circle of arcs), but I liked the name and the idea of infinity, so that's the pattern I always chose.

A relatively new treatment, EMDR's efficacy in relieving PTSD symptoms is impressive.

Sure, my back and neck were injured, but they took a back seat to the PTSD. Physical injuries were like a thousand-piece jigsaw puzzle. But there'd be a solution. We'd put pieces back together and see a whole picture, just like on the front of the cardboard box—there'd be an answer. But one evening at dinner a newscaster reported that a woman had died when the top floor of a parking ramp collapsed. I yelled, "No!" and covered my face. My fears could be validated. But my mind was like that puzzle—all corner pieces and no middle. How would I alleviate my paranoia, relentlessly morbid thoughts, fear of everything man-made? How would I again take for granted the stability of concrete and steel?

I try to explain the post-traumatic stress. Tense. Vigilant. I was genuinely afraid my mind would be my enemy, that I'd lose control of the ability to keep it together. Science and nature provided insights into the brain's workings and the role of fear in coping. "Animals that stay scared are more likely to stay alive," and, "Better to be scared and alive than happy and dead," a narrator of a nature show intoned. This was the natural inclination I fought, that to lose my fear would endanger me.

I go through a phase where I need to cry but can't. During waking hours I do anything but. At night a tear squeaks out of the corner of my eye, but that's all. I must cry until I can't anymore. But when I try, nothing. I can watch movies with the worst tragedies as if I'm stone. Broken, I need a therapist who'll help me access my tears. I must sob my life's worth and the knowledge that I'll face that death moment again. Will it blindside me like it did then? My future is waterlogged beneath the weight of all this salt. I'd transformed myself before; I'd do it again. It was with that hope alongside grief that I entrusted myself to a circle of healing, to scientific know-how. Just what my spirit required to go on.

It's a shame that there's a stigma about therapy, certainly in the general public but especially in the military (ptsd.about.com). My own misconceptions almost kept me from doing the work. Most significant, therapy taught me how to think differently. Whenever things happened, and I couldn't think of it in any other way, therapy offered insight. A way to open oneself that isn't feasible with family and friends. Therapy, when you're with a practitioner who's a match, can perform miracles. A safe place to look at trauma and examine any programming that makes us live a limited life. I see stigmas for the ruse that they are. A waste of time. I abhor them, their lack of utility and compassion. I'm lucky beyond measure to know this.

"How's the speed?" Barb asked. Without exception she asked me this. Eventually, I'd laugh at the question because my answer was always the same.

"The speed is fine, great."

With headphones on my ears and bean-shaped nuggets in my hands, we set off. At first the buzzing, beeping, and zooming were strange. I looked at Barb dumbfounded.

"Okay... what do I do now?"

She assured me: "It's okay. Just start talking. Let's start with the ramps. Describe the fear. The sensations."

I heard *buzz buzz* alternate left and right in my ears and felt *buzz buzz* alternate left and right in my palms and watched infinity circle round and round.

When I walk into parking ramps, I fight a flutter of sensation. Pressed onto my understanding, the unyielding quality of concrete. I feel how gravity presses me into the floor, and the soles of my feet feel a ghostly discomfort beneath the press of my body. When I look ahead, I notice the concrete extends far down my sight line, forty or fifty cars on one side, duplicated on the other, aisles and aisles of cars, five levels of structure performing a miraculous and underappreciated feat, silently holding invention aloft. Sounds jangle in my ears, through all of me—eyes, nose, bones, blood. Flutter, wisp, fleeting. Come. Go. Fight it, fight. Still, still, the physiological reexperience. Always self-talk. *You're okay. You've worked. You can separate the triggers.* Then, now, past, present. But still, you are here. *Blunt force injury.* No-no-no-no! Stop thinking about that! Moment to moment—all can change. Too heavy all this concrete. Too heavy. All these cars. The support columns. If the ramp floors pancake onto me, it will hurt. Blood, bone, belongs on the inside. Not breaching skin. Temper the self-talk with knowledge. Fight, fight. Mind, come back. Be safe. Be reassured. Put one foot in front of the other. Your car is around the corner down the outer edge. Keep going. Get there. Beep beep: press the unlock pad on the key fob. Open the door. Get in. Exhale. Put the car in reverse and back out of the stall. Circle the ramp. Fight, fight. Almost there. Circle around—circle around—circle around. Down—down—down. If the ramp pancakes onto you—if it does—your blood, bones, head, ears, eyes, nose, consciousness (blunt force injury), will hurt, will horrify. You won't live to tell about it or reexperience it. It will be done. There will be no more conversations like this in your head. *Stop it, fight.* Above, the ramp ceiling remains strong. *Almost there. You will not be crushed.* Turn on the radio. Let songs console you. Exit into the sunshine. You are almost out. You are out.

I heard *buzz buzz* alternate left and right in my ears and felt *buzz buzz* alternate left and right in my palms and watched infinity circle round and round.

In our sleep, during REM, the hemispheres of our brain exchange information like fleet-footed messengers. EMDR recreates or approximates what your brain does during REM. Based on the theory that bilateral stimulation of the brain helps human beings process events, EMDR is a bridge. A brain bridge, spanning the corpus callosum—the fibrous, membranous structure that connects our left and right hemispheres. This crucial communicator depends on our ability to traverse its flat white landscape like the concrete of a bridge deck. We continually cross the expanse. Start on one side. Arrive on the other. Repeat. Like a chorus, like poetry. When people experience trauma, they receive a message in one hemisphere. The message fires like an incessant child until you rectify what the brain is trying to resolve. EMDR helps a person resolve "stuck" messages. While you still experience triggers, you no longer automatically attach past experience to current reality. EMDR cuts or weakens the link.

It's an unbelievable sense of accomplishment to arrive at a place where you understand—to know that part of the problem is that the body remembers every trauma. And your brain presents the same problem until you figure it out. And when you wake up with a nightmare and think, I should be stronger than that, "should" doesn't change anything because you're not attacking the correct problem. Your body remembers. You have to marry up your body and your mind—and you can't do that, not quickly at least, without some help.

What helps is facing the truth and letting it be okay to have emotions. Alcoholism, drug use, violence, and abusive behavior are ills that started, perhaps, when people were taught to hide their emotions, equating crying with weakness. Simplistic? I don't know. But one thing's for sure, pretending the problem away doesn't work. Acknowledging my pain and asking for help gives me the power to change course.

120

The narrator of *The Crumbling of America* said: "Bridges have, of course, failed before in America. Miscalculations in wind dynamics brought the Tacoma Narrows Bridge down in 1940. The Silver Bridge in Ohio killed fifty-seven when it collapsed in 1967 due to a defective link in a suspension chain. More recently, the Mianus River Bridge collapsed in Connecticut in 1983, killing seven. Poor maintenance and inspection allowed rusty pins that held a central section of the bridge in place, to fail. Coming in the wake of Katrina and the growing sense of unease about infrastructure in America, the I-35 collapse was deeply disturbing."

121

An interfaith prayer and memorial service at the Basilica of Saint Mary in Minneapolis marked the one-year anniversary of the collapse. Sober and thoughtful, the service paid tribute to the faith and culture of each person killed. With each note, sound reverberated off the marble sanctuary's walls and flounced through our bones. Buddhist monks chanted, full of shocking power, the raw tide of the sound and rhythm. An Islamic student chanted for mother and daughter victims Sadiya and Hana Sahal. Native Americans chanted and beat a drum for Julia Blackhawk. Indian folklore spoke of beating the drum loudly so the dead could hear. I wondered if Julia could.

Afterward Rachel and I rushed to get to the procession, the walk to the Stone Arch Bridge, a bridge upstream from the 35W collapse site. We found a meter, parked, and walked the neatly kept sidewalks toward the new Guthrie Theater, then through the open area, down steps to the River Road.

The procession began to file past us within seconds of our arrival. First, fire trucks and ambulances with their lights spinning but sirens silenced, coasted past. Then came the sounds of dozens of bagpipers. Officials from the State of Minnesota were closely behind. And how strange how far I've come this past year, to watch these people file by. I was able to name them one by one and associate a memory of days at the Capitol with them. Governor Pawlenty, Senator Larry Pogemiller, Speaker Anderson Kelliher, Representative Winkler (who sponsored the House version), Senators Dibble, Carlson, Klobuchar, Ellison, Coleman, Saltzman, and so on. Rachel and I stood on the steps watching, my fear of homophobia inconsequential at that moment, my hand laced tightly within hers. We shifted and settled our breathing, and I looped my left arm around Rachel's shoulder. She wrapped hers around mine. And the feeling was nice. A breeze, some shade, and us standing there. Together. People filed by, including a woman with many gigantic cameras hanging from her neck. Out of the corner of my eye, I saw her point them at us, and I heard the rapid clicks of the shutter. I wondered what that photo would show or where it would appear. Next we saw other survivors, Dave and Deb. They waved, and I waved back, grabbed Rach's hand, and we dashed into the procession.

Then, at 6:05 p.m., there was a moment of silence, and Mike Martin, a Minneapolis police officer, released a dove for each of the victims. With each dove he read a name. They flew over our heads, white flashes like spirits evaporating. The media walked the sidewalks with survivors, families, and friends. Ambulance lights spun, and helicopter blades chopped: sights and sounds reminiscent of the day the bridge fell.

122

"Congress also must ensure funds sent to states for bridge repair are used only for that purpose. Today, states can transfer up to 50 percent of their bridge funds to other purposes—even if they have bridges clearly in need of repair."
Transportation for America website, "The Fix We're In For: The State of Our Bridges," www.t4america.org

123

That evening Rachel and I went to a bar near the bridge with friends, when we met a construction worker whom I'll call Ben. We began talking about the collapse and the anniversary. Ben had a rapport with his buddies and seemed at home in the sports bar. He said, "MN/DOT had a $2 million contract to repair that bridge, and they didn't go with it because of the money." Yes. I remembered reading about this—a legislative commission concluded that money concerns had influenced some of MN/DOT's decision making. He confided things I would have never known, things I'd never know from the media. A part of me didn't want to hear it. It had been a taxing day. I wanted to believe that things would get better, that the long stressful year could be relegated to memory and worshipful remembrance.

But Ben kept talking. "MN/DOT is well funded, but their approach is mismanaged. They get out on a job, and they're instructed to replace the bearings or some other repair on a bridge. They go out to perform a repair and then find extensive decay. The contracting company tells its workers that MN/DOT wants to get ten more years out of a bridge." He said it's known between construction workers that the Cayuga Bridge (a bridge that heads north out of downtown St. Paul) is in rough shape. Workers won't cross it and plan any route around it. Things haven't changed much, even after all that's happened. About a bridge they were working on in New Ulm: "If a Geo Metro were to run into the guardrail, the rail would just fall apart—it was so rotten." But, Ben said, "we couldn't fix it because it still looked good on the surface, and a repair wasn't in the contract." This situation occurs over and over again. He put his head into his hands and shook his head. He continued. Near where the city stores its garbage cans, there was a pier under an overpass, the I-394 freeway that leads west out of downtown Minneapolis, that was held up with straps. All the salt and chemicals pour on the bridge deck and over time drip down and corrode the cement and rot the piers.

"Why are you telling me all this? Now I'm upset."

"I'm not trying to upset you. I'm just telling you the facts."

"The pier that's held up by straps—I want to find where it is."

"The public doesn't know about it. They can't see it. You can only see it from the Dunwoody parking lot." Dunwoody College is a school of technology situated where the Interstate freeways 394 and 94 intersect, west of downtown Minneapolis (my regular route to work, where I had noticed that people parked beneath the roadway).

Ben said it's a mind-set, a pattern. "We get out there [on a job] and begin working, and when we start jackhammering, a bunch of concrete starts falling off, and it's a ticking time bomb out there. They find problems but can't fix them. [Commissioner of Transportation] Molnau said we're going to do the same job for half the money. So they picked and chose what to fix. 'Deferred maintenance.' Only do what they have to do and defer the rest. Workers come back and say we want to do this [repair] and they're told, No, we need to defer that until later, which basically means we'll ignore it."

"You have to tell someone," I urged.

He said he couldn't, something about trust among the crew, fear of reprisal.

"But how can this continue?"

"I don't know, but I can't be involved."

124

The narrator of *The Crumbling of America* said: "The American Society of Civil Engineers issues an infrastructure report card every three years for fifteen infrastructure categories including bridges (C), dams (D), drinking water (D-), levees (D-), roads (D-), wastewater (D-), the grid (D+), schools (D), solid waste (C+), transit (D), inland waterways (D-). America's current grade point average: a D."

125

I now thought of life in terms of energy: how much and what kind. I weighed the worth of X, aware that time flowed like rivers beneath bridges. The lock and dam slowly released water like our days that slipped from awareness on a current. We couldn't stop it. Constantly in motion, life was awareness. Going to sleep, I was aware of space. My body, physically, was healing—able to be in bed without acute pain. But two toes on my left foot rubbed wrong in my shoe. Trying to fall asleep, I often wiggled those toes apart. When my bones rubbed together, the sensation stayed with me. That little toe jammed into the claustrophobic one, made me antsy. My stomach restless, I wanted to move and shake it off. I recognized the flight response. Fight or flight. This simple act got the thoughts moving again, thoughts that lived subcutaneously. I worked through them as if through a scavenger hunt. Collect a thimble, get a keychain, find a hat, a ball, a dollar bill, a leaf, a pen. The route was a subconscious exercise in survival, to find my way past that annihilation, the uncomfortable knowledge the fall gave. It flared in the silence of presleep, when our sheets warmed to our bodies and our heads fit into the groove of our chiropractic pillows.

Energy. Physiological conscious awareness. I had a handful of tools to fight the fear. I soothed flight away. Come to the present, to bedtime, safety, to dreams and letting go. But I didn't sleep well at all after what I had heard the previous night from Ben.

The morning after the one-year anniversary—August 2, 2008—I left home at dawn. For my birthday Rachel had surprised me with tuition for a three-day writing workshop at St. John's University in Collegeville, Minnesota, a few hours northwest of the Cities.

Though I was dead tired, my mind spun. *It doesn't have to be inevitable... another tragedy like this.... They can change.... Good people working on our bridges... they've just been operating for too long under an old system.... These bridges were built in the Cold War era, and now they're ticking time bombs.... With the salt and the chemicals that we use in the winter, especially in the city.... They're aging, and now we have to play catch-up.... We have to fix them now.... We shouldn't build them if we can't keep them maintained.... Better to not build a bridge that will crumble than build bridges and let them break.*

After the first class, attempting to settle into the spartan living facilities, I couldn't bear to be there. It was wrenching to be away from home, but I also felt lost and unsettled—more than being homesick. I came home the same day.

The words of Ben's wife came to me: "I don't want my husband dying [while working] on a bridge."

What could I do? I heard my mind answer. Find them. See them with your own eyes; take photos.

126

In the car with Rachel, I said, "Oh no oh no, why?" as I steered the car down the lovely curve of a high and newly built overpass. I was thinking grand and cosmic thoughts about fate.

"Why do I know this now? Why do I know this information? Oh god... what's this next year going to be?" Rachel chimed in, concerned that I'd try to tackle another huge problem instead of investing in my healing. "You know there are people whose job it is to do this. They have families too. They drive these roads too. There will be people looking out for this, working on this." I kept quiet.

I looked her way for a moment and then quickly returned my gaze to the task at hand, driving. How I wanted to believe her, to be able to let go and focus on healing and peace. I mustered a halfhearted nod.

127

James Meigs, executive editor of *Popular Mechanics*, said in *The Crumbling of America*: "Bridges fail two ways. One is with a bang. But much more commonly, they die with a whimper. They become degraded. The bridge can't perform the way it's designed to perform, and that's what we see happening in a lot of parts of the country."

128

On August 17, 2008, at 9:00 a.m., I left home to find that decaying guardrail in New Ulm, a town about two hours southwest of the Cities.

I was driving, and there was construction over 35W, a gap in an overpass ahead of me. It was like a miniature version of the big brand-new bridge (35W), where they're lifting up segments, and the segment that they're lifting up has the opening in it, and it looks just like the bridge. There's a gap there, and as soon as I saw the gap, I don't know why, but I panicked for a second.

In our Western society we call intuition hocus-pocus, but it's instinct that leads an investigator down a shaded corner, an unplanned path. At the same time I was filled with doubt. I didn't know what I was doing. Was this stupid of me, to even think, Oh, l'il ol' me—I'm just going to hop in my little car and take a little drive and, lookee there, find this guardrail, and then what? Shake this guardrail? I couldn't exactly drive my car into it to see if it crumbled like Ben said. Assuming I found it, what if construction crews were working and I couldn't get close? I tried to get ahold of workers to ask some of these questions, but I couldn't get ahold of them. Could I even walk on that bridge? A lot of unknowns, but I just had to do it, so I drove two hours (each way) just to see if I could find anything. I didn't know where I was going exactly, but I was following the directions. Going on intuition, trusting it—riding on County Road 8.

Listening to Bruce Springsteen's greatest hits, I drove, blasted the music, and tried to enjoy the ride. At 10:00 a.m., viewing the open country fields, I jotted on a notepad: "There's something really peaceful about all this corn ... all these tall stalks waving in the wind. There's just so much corn out here."

Luckily, on a Sunday there were no construction workers. On one side, where it was closed, I had to go all the way around, and then I ended up going the wrong way a few times. Then I finally found it. I drove around in circles, trying to figure out where to leave the car because the bridge was closed. There was no ideal place, so I parked askew near a cloverleaf exit. I walked over. The heat beat down, and I walked quickly to the deafening sound of insects on a deserted bridge. Being out there alone gave me the creeps. There was no pedestrian walkway and no shoulder, so I walked quickly along the length of the rail looking for the decay. Found it! I pushed on, took my pictures, and hoped nobody would see me and ask what I was doing.

When I got back home that afternoon, Rach and I piled into the car to check out the Cayuga Bridge's deck underside.

129

The narrator of *The Crumbling of America* said: "From the 1930s to the 1960s, the United States built the greatest infrastructure the world had ever seen. And it played a critical role in our rise as a superpower. That whole build-out is reaching the end of its life cycle and groaning under loads it was never supposed to handle. Design flaws, not enough money, normal corrosion, and decades of deferred maintenance have conspired to break America's infrastructure down."

130

So spooky being beneath bridges. They are desolate places, no people, the constant din of overhead traffic, yet eerie the areas overgrown with weeds, dirt in areas where grass won't grow. There is gravel, and the ground is uneven. There are empty soda cans and liquor bottles. This is not a place I have ever voluntarily visited, let alone wanted to visit. I feel a vague sense of anxiety, but I press forward.

Starting with the underside of the deck of the Cayuga Bridge—a 1,285-foot span of I-35E north of downtown St. Paul, one that's been on replacement lists for years. With a sufficiency rating of 40.8 on the 100-point scale, it carried nearly 150,000 vehicles a day and was one, Ben said, that those in the industry "won't drive on until it's replaced." I would find those places on a hunch. The heart knew when something wasn't right. Facts told only part of the story.

Beneath Cayuga I looked at the deck underside with wide eyes. The vastness of the space beneath the bridge surprised me for some reason, though it probably shouldn't have. From ground to "ceiling" it was cavernous; the eye followed far and wide. When cars passed overhead, their tires dully thunked, jangling and echoing through the steel. Aware of my post-collapse paranoia, I rushed to look. I snapped a series of pictures that captured the extent of time, deterioration, and neglect. On the deck's underside I saw cracks, blisters, and tree branches wedged between beams. In one picture light shone through a hole in the deck.

131

Dan McNichol, author of *The Roads That Built America*, said in *The Crumbling of America*: "Structurally deficient could mean something like, 'It needs to be replaced in five years,' or structurally deficient could mean, 'It's on the verge of a collapse.'"

132

I compiled my photos, totaling thirty-four pages, and sent them in a "Dear MN/DOT" letter.

AUGUST 19, 2008

Minnesota Department of Transportation
Central Office
Transportation Building
395 John Ireland Boulevard

Saint Paul, MN 55155
Email: info@dot.state.mn.us

Dear MN/DOT,

Like so many others on August 1st of 2007, I was almost killed on the old 35W Bridge. As someone who survived, I talk to people—and people talk to me.

They say there are widespread, systemic problems at MN/DOT. That they're well-funded but mismanaged. That the way they're running things, it's going to happen again—guaranteed. I didn't want to believe it.

So I started asking questions—lots of them. I learned more than I wanted to know: guardrails are allowed to rot; piers are held up with straps; workers find problems but aren't allowed to fix them because it still "looks good" and MN/DOT wants to get 10 more years out of a bridge. I still didn't want to believe it.

So, I went to see for myself.

14. Pier held up with straps; I-394 facing Dunwoody College. © 2008 Kimberly J. Brown.

I want to know: Why are you allowing this deterioration?

Example #1: Pier Held Up with Straps

A pier under 394 near Dunwoody is rotting. A huge piece of concrete had been held in with straps. The straps are gone and in their place is a pile of broken concrete. Minnesotans have seen enough broken concrete.

Example #2: No Repairs for Rotting Guardrail

Workers were sent to Hwy 15 to do deck repairs and painting on Bridge 9200, which crosses the Minnesota River in New Ulm. When workers found that the guardrail was rotting badly, they were told not to fix it per MN/DOT's wishes, which were to "stick to the contract" and the guardrail still "looked good." I'm told that if a Geo Metro crashed into the rail, it would crumble, it's so rotten.

Alongside the rail, down low, I saw corroded rebar and crumbling concrete.

Example #3: Cheap Fix for Cayuga Bridge

This bridge was ready to fall apart five years ago when it was worked on. Instead of repairing the deck properly, MN/DOT opted for a cheap

15. No repairs for rotting guardrail; Hwy. 15 Bridge 9200, New Ulm, Minnesota. © 2008 Kimberly J. Brown.

fix. Workers milled off the surface and overlaid the deck "dirty." To give you a better idea, they haven't driven across it since and have no plans on ever driving on it until it's properly replaced.

Cayuga is rated 40.8 and carries 150,000 cars per day. Dan Dorgan of MN/DOT says that it isn't "fracture critical." If we compare the bridge rating to grades we give our children in school, a score of 40 out of 100 would get an F, and it's still 20% short of a D-.

I went to the 1,200-ft.-long bridge and looked at the underside. Corrosion and rotting concrete are widespread on the guardrails and support structure. But once I got underneath, I saw signs of decay from cracks to rips, topped off by places where light was shining through holes in the deck.

So I ask again: Why are you allowing this deterioration?

Your Response

In your response, please don't say our bridges are safe. After the bridge collapse and other close calls, I just don't believe you.

A 1,200-pound piece of concrete falling from the 35E/Maryland overpass (3) and drain pipes over 35W crashing through windshields (4)—by their very nature—are not indicative of safe structures.

16. Cheap fix for Cayuga Bridge; St. Paul, Minnesota. © 2008 Kimberly J. Brown.

MN/DOT spokesperson Kent Barnard downplays the potential danger. As reported by MPR: "Chunks of concrete have fallen off of other Minnesota bridges in recent years. Barnard says last year a piece of concrete fell off of the Highway 7 Bridge onto Highway 100 in St. Louis Park." (2) Really, this kind of thing just happens?

Riding my bike recently along Minnehaha Pkwy, I spotted a basketball-sized chunk of concrete on the ground just off the path. It had fallen from the Nicollet Bridge above. Mr. Barnard says we're safe.

Meanwhile, MN/DOT continues work on its image (6), telling the public we're safe. And of course we want to believe you. But Minnesotans deserve better. Those of us who fell with the old bridge know too well that "image" won't keep us safe.

Don't say "It's just cosmetic." I checked with an engineer and it's not.

For example, regarding the pier: An engineer tells me this probably isn't immediately dangerous, but it's certainly not normal. The pier is definitely weakened and it looks as though the beam has stress cracks that could accelerate the deterioration as moisture and salt enter the beam. The discoloration of the beam and the rusty reinforcing rod is evidence of salty water damage. When the steel rusts, it expands and breaks off the concrete and makes mini-fissures where more water can enter (spalled concrete).

Corrective measures that should have been taken:

1. Determine how or why the joint is leaking saltwater onto the beam and pier.

2. Repair the joint to stop the saturation.

3. Inspect the pier and beam to determine the extent of damage, depth of cracking and spalled concrete. Damage can range from cosmetic to structurally unsound (and anywhere in between).

4. Determine the best type of repair to restore the load capacity and cosmetic protection of the pier.

Don't say "Maintenance is scheduled for the future." This "deferred maintenance" is just another way of saying "repairs ignored."

Under the previous transportation commissioner's tenure, MN/DOT was under pressure to cut its budget significantly (allegedly by 15%), choosing to "defer maintenance."

17. Another rotting pier under I-394. © 2008 Kimberly J. Brown.

An engineer says that preventive and restorative maintenance MUST be done on concrete structures without delays. Once steel gets hit with moisture and begins to corrode (and expand), it begins to stress the concrete and crack it.

"The collapse of the I-35W Bridge should be a call to action," Sen. Amy Klobuchar said in a statement. (7)

The deterioration of this pier should have been corrected when it was first seen in order to slow the deterioration. In two more years the concrete will have broken enough that it isn't safe and much more difficult to repair.

This pier is nearby and is beginning the process.

Don't say "It's because of money." MN/DOT says that money never influences safety decisions. (5)

MN/DOT opted for a cheaper fix, instead of going with a nearly $2 million contract to reinforce the steel on 35W. (5) And look what happened. Ignoring repairs to overpasses, bridges and roads doesn't address slow deterioration from traffic, weather, salt and chemicals, thereby making these structures ticking time bombs. This neglect cannot keep happening.

Repair Now

I didn't want to believe it, but the proof is there, just like I was told. Top leaders at MN/DOT need to do an about-face. How can you keep track of this stuff and not take action?

While some effort has been made (like the closure of problem bridges), the examples above are evidence that MN/DOT needs a new attitude. I would rather hear, "We've got problems. Yes, we do. But here's how we're going to handle it."

Start focusing on substance and your image problem will take care of itself. When we get in our cars and drive every day, we shouldn't have to worry about the structures we depend on. Bridges shouldn't collapse and overpasses shouldn't crumble.

What other maintenance is MN/DOT deferring that the public would never know about?

I beg of you—stop planning for repairs later. Start fixing problems now.

Sincerely,
Kimberly Brown
35W Bridge Collapse Survivor
In Car B3
Minneapolis Resident

P.S. I attached additional photos so you can see more of what I saw.

1. Associated Press, NPR—"Little Progress since Bridge Collapse," 7/31/2008
2. minnesota.publicradio.org—"Lawmakers Want Tougher Bridge Regulations," 7/28/2008
3. wcco.com—"Story of Car Hit by Falling Concrete," 7/27/2008
4. wcco.com—"Saved by Inches: Pipe Falls on Car from Overpass," 7/29/2008
5. StarTribune.com—"Phone Call Puts Brakes on Bridge Repair," 8/18/2007
6. TwinCities.com—"Rebuilding Trust at MnDOT, One Year after the Bridge Collapse," 7/27/2008
7. TwinCities.com—"Did Engineers Seal Fate of I-35W Bridge?" 1/16/2008

133

The narrator of *The Crumbling of America* said: "A big reason is age. Most modern bridges in the U.S. were designed to last fifty to seventy-five years. The average age of bridges today is now forty-three, and a lot of the concrete and steel that went into them is breaking down and rotting away. Bridges need basic maintenance and repair: paint jobs, patch-ups, clearing away ice and salt, all help to keep corrosion at bay. Unchecked, rust can spread like a cancer and take a bridge down. But a great deal of bridge maintenance in America has been put off for years, even decades."

134

A week later on August 26, 2008, I sent my "Dear MN/DOT" letter again. On September 2, 2008, I resent it a third time. Three weeks passed with no word. Frustrated, I sent it to various media outlets. The *Star Tribune* wanted to run a story about my findings, and then the paper abruptly backed out. A state senator I met during the compensation process hypothesized that it was probably a veiled refusal to be critical of Governor Pawlenty, who had been rumored to be on the short list of potential VPs for then-presidential hopeful John McCain. We were in the swell of DNC and RNC political convention coverage, so the timing was unfortunate. I sent it to other outlets, and they wouldn't touch it either.

I hung six bridge maps on my office wall. This is what I saw: three close-ups of the Minneapolis and St. Paul metro area and three exploded versions of the state—wide-angle views of the puzzle. Divided shapes of counties ripe for jigsawing. Wright County interlocked with Sherburne. Stearns with Benton, Carver with Hennepin. St. Louis, Lake, and Cook Counties, vertical counties on the arrowhead like doors. Imposed upon the ordered puzzle of these maps, stars and triangles symbolized the bridges we've birthed as a people. All these bridges . . .

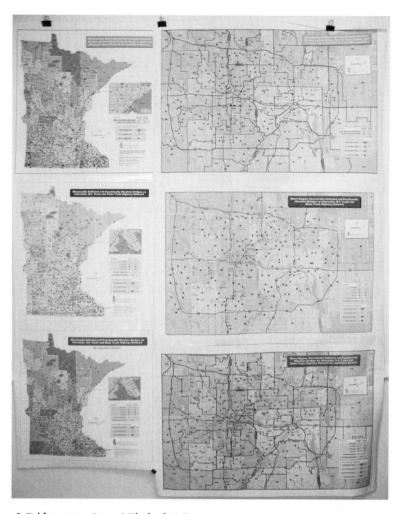

18. Bridge maps. © 2008 Kimberly J. Brown.

We allow these engineering feats, once new and magnificent, to age unaided. Inspection only, deferred maintenance. We measure progress on an inspection report like a nurse notes a patient's pulse, temp, heart rate. But we say our hands are tied—budget cuts. This is why we don't fix them. We build new instead of repairing the old. Shortsighted, we're lured by bigger, better, shinier. The old bridges corrode. Blistered steel spreads like a rash, multiplying beneath our harried pace. Zoom-zoom as we zip across the expanse. We have other concerns.

On the exploded map view: metro rivers, blue dots for ponds. Ambling freeway with a green wiggly line, ventricles pointed outward, around rivers and topography. Splotchy spots like rosacea. Arteries and veins of a collective body. Colorful, complicated, expansive, vital, and in perpetual motion. The icons plotted represent all the opportunities, the choices we have. Will we elect people who promise to cut taxes, the message in line with the political flavor of the day? Or will we choose to care for a declining system?

I had crazy dreams. I got lost. I found my sister at a restaurant, but we were blocked in. A fellow bridge collapse survivor who'd been injured badly couldn't get by. We had been sitting on a large bench in the middle of the room. We stood, plying servers to our aid. Then I dreamed about seeing an email displayed on a screen. A woman read from it, "On a bridge I feel like a fish out of water." It was from a bridge collapse survivor, and the email stamp said Melbourne, Australia. I saw our planet from space, the whole oceanic orb, floating as if held by a fixative, but the clouds frothed, and the blue churned. I saw the planet, land covered in sea, in pulsing fathoms. The same liquid that swallowed the 35W victims. Then the planet spun.

In reality no survivors hailed from Australia, my mind manufactured it, but I felt adrenaline in my chest. I woke Rachel up. I said I knew it was silly, I'm a grown woman, but I'm freaked out and can't get rid of the feeling. We readjusted, moved toward the center ridge of our Sleep Number bed. Rachel pulled me close. We lined up like Legos. She reached her arm around my ribs, and I clasped her hand. A skein of light shone through the window.

Rachel asked, "Are you still scared?"

"Yeah."

"Can I talk you down?"

"Yeah," I said, feeling like a little kid.

She began, "First, they're from Australia."

I laughed, drew the covers toward us, under my chin.

"Second, you don't need to worry about them. They can take care of themselves."

I stared into the charcoal night.

"Third, you don't have to take care of the whole earth."

To this I sighed all the air out of my lungs, and then she nuzzled and kissed the back of my neck. Voice muffled, she whispered, "You're. O. Kay." I sucked in breath. We squeezed. Over the back of her hand, I ran my thumb in short arcs.

135

The narrator of *The Crumbling of America* said: "According to civil engineers, we're in such a hole that it'll take $2.2 trillion over the next five years just to bring America's infrastructure to an acceptable level. But only 9 percent, or 72 billion of the $800 billion stimulus package passed by Congress in February 2009, was explicitly allocated for infrastructure projects."

136

My back and neck are sore. The changing of the seasons makes my post-bridge body hurt. This is the time bridge survivors start hurting more. At group, survivors commented on how cranky our bodies were with the changing atmosphere, the evaporation of summer's humidity and the shift into fall's crisp coolness. It's like our bodies gauged the shift from season to season in cartilage, soft tissue, and vertebras that stick upon themselves.

My head feels like a bowling ball on my neck, perched unsteadily, making me a real-live human bobblehead. My back is tender like chicken. I toss and turn; no position is comfortable. At night, when Rachel says, "I'm going to go to sleep now," I get up. I must grow exhausted before my body will submit to the gravity of horizontal rest, pressing on the sore spots that remember.

137

In *The Crumbling of America* Andrew Herrmann, Report Card advisor, American Society of Civil Engineers (ASCE), said: "All of the categories rated by ASCE for the infrastructure are affected by corrosion. It's lack of maintenance that shortens the life of infrastructure."

138

The bridge just never goes away. I don't think I can ever be happy about this new bridge. Others can be happy about it.

Rach said, I know... I feel no happiness about it whatsoever, and I won't. If it pleases others, bully for them. But they'd better not expect that reaction from us and the hundreds of others affected by the collapse. To me it's just a replacement for something that killed thirteen people, almost killed you, and injured and traumatized many, many others. That's all it is. Period.

I just didn't realize how close the new bridge was to opening. I hadn't decided how I felt. I think I know that it's okay for me to never be "happy" about it.

Meanwhile, we still have forty-two bridges in the state that are rated lower than the old collapsed bridge, and there are repairs in the system that are ignored year after year. (I still hadn't heard back from MN/DOT in response to my letter but was doing interviews regarding my findings.) I understand that others will be happy not to be inconvenienced anymore.

I recently met a person who was involved with the new bridge, and he said to me: "You have to be there to walk across it when it opens! Make it come full circle!" He smiled. I felt such a feeling of revulsion hearing that pronouncement.

That spot, where the new bridge crosses the river, was a graveyard for thirteen innocent people. I'm always acutely aware that one of those people could easily have been me. Others can be happy about it.

Forget for a moment what you've read in papers or seen on TV. I continue to feel (and may always feel) psychic pain. "The bridge" never seems to go away. If I'm not thinking about those poor people's last moments, I'm thinking about what might have happened, and my body wants to hurt for them.

139

The narrator of *The Crumbling of America* said: "Corrosion is the number one killer of bridges in the U.S. Wind, water, and salt combine to create acids in steel supports and exposed rebar that will eventually eat away the steel and bring down a bridge. That's why basic maintenance like painting is critical. Paint forms a seal against the elements and protects the steel."

140

The *Star Tribune* wanted to make my "Dear MN/DOT" letter a personal profile on me. The reporter who'd potentially cover the story and I corresponded.

> The profile idea . . . I just really don't want to be treated badly, like whistleblowers often are. Whoever is being brought to task finds ways to say negative things about the person who is the messenger. Then the message loses focus and people miss the message. I do not for a second, if I can prevent it, want people saying or thinking I'm doing this for attention. If I wanted attention, there are easier and much more fun ways I could get it.
>
> People at MN/DOT and general citizens shouldn't truly be focusing on me. They should be focusing on the fact that things like a pier that supports (how many) pounds of weight have been deteriorating (and never repaired). (I pulled inspection reports—year after year, no repairs.)
>
> This is serious stuff and I want serious attention brought to it. As much as praise feels good, I don't want adulation for this. It's too important. If the "profile" can explain my feelings about that, I'd be much more up for it. (I think).

> Gotcha. Let me talk to my editors about this . . . maybe the story is a profile about how you don't want the issue to just fade, like issues often do . . . you're on a mission to get things changed.

> Exactly. Thank you.

141

In the television documentary *Modern Marvels* the narrator said: "Corrosion costs the U.S. economy $300 billion ($300,000,000,000) each year. That's four percent of our gross domestic product. Nearly half of what we spend on foreign oil. Or roughly a thousand dollars per American citizen annually. Corrosion's destructive power can be found everywhere. From rusting railroad trestles to deteriorating oil pipelines. But what exactly causes this scourge that eats away at our society?"

142

After trying to find a media outlet to share the information I'd gathered and assist with MN/DOT's unresponsiveness, I was elated when I found a home for it. In September 2008 a nonpartisan think tank called Minnesota 2020 expressed interest in the evidence I had gathered. Maybe now, partnered with a research-based organization like Minnesota 2020, which had published a veritable library of articles featuring inadequate maintenance, maybe now our dangerous infrastructure would get the attention it deserved.

But I had loads of work to do. I had to write another speech and think about the message I would convey, think about the ways that reporters—well intentioned or not—might throw my message off track. The new bridge was nearing completion and would open soon. Would the people who live and work in the area be so awash in positivity, in wanting to move on, that they would forget?

The think tank would publish my evidence and findings on its website, send the information to its mailing list of fifty thousand subscribers, and stand with me next to the rotting pier for a press conference.

I created an email address, survivor35w@gmail.com, to give MN2020 subscribers a way to contact me if desired. Since the news media only has time to show a nanosecond of the press conference, I wanted the transportation committee to see the full text of what I said.

I would speak clearly to the cameras, stifling nerves, channeling the quiet pulse of those whose voices had been silenced.

143

Florian Mansfeld, professor of engineering, University of Southern California, said in the television documentary *Modern Marvels*: "Many different environments can cause corrosion. A corrosive environment can be seawater, can be the atmosphere, can be a chemical solution, an acid."

144

The narrator of the television documentary *Modern Marvels* said: "It's a natural electro-chemical reaction, occurring on an atomic level in most metals. Water conducts electricity that allows the exchange of electrons on the surface of the metal. Positively charged atomic particles called ions flow from the metal and combine with the oxygen in the water to form an oxide. In the case of iron, that oxide is rust. Iron is found as an ore. Iron oxide. We put energy into the ore to produce pure iron. When this iron is exposed to a corrosive environment, it rusts, releasing that stored energy and reverting to an oxide state. Different metals corrode at different rates. Reactive metals like iron oxidize easily. While more passive metals like gold resist corrosion, earning them the title noble metals. That's why you'll find non-oxidizing metals such as platinum used in medical implants. But in a world with an infrastructure built primarily of iron and steel, corrosion is rampant and its impact can be catastrophic."

145

Minnesota 2020 scheduled the infrastructure press conference for Monday, September 15, 2008. Conrad deFiebre, a Minnesota 2020 fellow, produced illuminating facts, and standing with him and other researchers added to my credibility. All four local TV news channels were there, plus a *Star Tribune* reporter. The press conference went so well that Rachel and I were on a high for hours. We were, anyway, until we saw the *Star Tribune*'s early online edition. Throughout the piece the reporter used loaded and inaccurate language about my intentions. The reporter wrote, "Kimberly Brown showed off an eroded pillar on I-394 and asked transportation officials and legislators to fix bridges now before they become the next to collapse." Then he made a potentially devastating misquote: "Memorials and ceremonies to honor victims and survivors are good, but we have to move on." His comment intimated that I rejected the memory of the victims and lacked empathy. What I had actually said was, "Creating a memorial and having ceremonies is good, but we need to go further." I was shaking when I read the article. I had worked so hard with Minnesota 2020 in the last week to create a nonpartisan, credible, inspirational presentation—one that would get citizens' and politicians' attention and hold it long enough for people to effect some positive change in our neglected transportation system. But instead of reporting the context and facts, it was like this reporter hadn't even been at the press conference.

Luckily, the online version was an early version of tomorrow's paper that had yet to go to print. Frantic, I called the *Star Tribune* and insisted that the most dangerous mistakes be fixed.

What made the newspaper experience even more unpleasant were the people who posted mean-spirited comments in response. One reader's comment said, "Does [Kimberly] make a living inspecting bridges now?"

Why, no. Unpaid. Not a dime in income earned, working outside of my industry, 100 percent volunteer.

Another anonymous commenter responded: "Why is this woman the subject of local newspaper coverage? She's not just a victim who's 'concerned' that the state's bridges are in disrepair. She's a tool of a democrat party group that's out there criticizing the governor for not simultaneously repairing all the roads and brideges [*sic*] in the state. Her phoney 'concerns' are not newsworthy."

Against good advice I responded to comments with "check the facts" responses. Taking pot shots from strangers fueled the futility I felt. When a citizen goes the extra mile and it's met with ignorance and disdain, how can anything be fixed?

After these battles, because that's how it felt after all of the work and energy I'd put into it, I'd collapse onto the couch—either so numb, depressed, exhausted, or disheartened—that my only desire was to watch television until I could feel nothing. Absolutely nothing. Or I'd be so completely overwhelmed with grief that the tears would come in torrents, hard and steady, and I'd fall into Rachel's arms. It was a horrible, difficult time.

But I also received positive emails. Many expressed appreciation for the information and concern about our infrastructure—but one chilled me with its prophecy.

Former state senator and gubernatorial candidate Becky Lourey wrote:

Dear Kimberly: I can understand your frustration. Having this knowledge is a burden. When I was serving Senate District 8 in the Minnesota Senate, a person high in Minnesota's Department of Transportation came to me. He started talking normally at first, but then, overcome with the burden of his knowledge, he put his head in his hands, and said through a voice full of emotion, "Our bridges are falling apart and our highways are deteriorating, and people are going to be hurt . . . and all I am allowed to say to you is, here is our schedule this coming year for the district you serve."

146

In the 1970s cars rusted. You might remember products like Rust-Oleum that promised to "eat rust." It promised to protect metal from rust and corrosion by blocking water and moisture. This was before metals were dipped into a bath, a solution that sealed the surface from exchanging atoms with the surrounding air or vapor or liquid. Metals aren't static, as they seem. An invisible process takes place that breaks down the metal. It's a slow process, invisible to the naked eye, like plant or grass growth. Slower than grass grows probably. But a process that moves in one direction toward disrepair. It will not reverse itself on its own—only with human intervention, the concerted effort of preventive maintenance. That 1970s car owner who invested in a can of Rust-Oleum, a coat to keep that cherry red Camaro cherry red.

147

Stephen Flynn, author of *The Edge of Disaster*, said in *The Crumbling of America*: "We Americans today have been a lot like the generation who's inherited our grandparents' mansion. People drive by going, 'Nice house.' But when you go inside it, it's falling apart."

148

Politicians are stewards of sound bridges and other systems as well: of the earth and sky and water. Like pipelines that deliver clean, safe natural gas to our homes and businesses so that we may stay warm and healthy in the long winter months. If only it weren't up to them to determine how to protect the ecosystems and man-made systems we build, that sustain us, that move us across distances so we may remain connected. Then maybe the answer would be clearer.

We should complain less about taxes and remember what they do. They protect us. They build futures. They keep bridges and dams and parks and stadiums—monoliths of our identity—from crumbling beneath the weight of daily demands. Who among us has ever experienced a personal failure because he or she contributed too much? I'd give more to bring those thirteen people back, to see the unappreciated workhorses dotting our cities and towns whole, gleaming, strong, and sustained. It is there we must focus. The future isn't just in the new. It is rooted in the humble, in history.

The future isn't only in the flash of a brand-new overpass that metes our pride but in the sheen of the undersides, the under-seen invisibles of bridges, the thousands that decay as I write these words. It is rooted in invisibles. It is rooted in invisible forces like physics in which members are in tension, being pulled apart—in compression, where weight presses down—and shear, in which opposing forces slide past one another, trying to make way. The system that moans as we build shrines to excess to exalt the talented few. It is not there that we should build our future but in the old, where we know what has passed.

One winter I sprinkled salt on the icy back steps of my house. Later the ice was gone but in its wake, a pea-sized hole. A blemish magnified a millionfold in the structures on which we depend. It's not that we don't know how. It's not like diabetes, cancer, or AIDS, for which there is no known cure. Winter brings hardship to people, requiring our vigilance and attention. Underneath the salted roadways that soar over land few tread upon, salt corrodes steel. Eats it like dinner. From bolts to bricks, mortar to rebar, concrete to cable stays, they were raised by humans.

They must be sustained by humans. Without repairs they become a back step pocked with acne-strewn holes. There's only one option if we ignore the underside of these bridges, when we defer maintenance. Only one way this story ends when we fail to heed the collective's warnings. And that way is down.

Fair is fixing bridges. It takes money. It takes people in charge to recognize the system they are part of, to cease half-truths, to cease the proliferation of the simplistic view that our taxes pay for excess. If that were the case, wouldn't we have an opposite problem? "All of our bridges are sound! Waste!" I can dream of an opposite reality: "Those bridges can withstand a margin of decay! They don't need that spending spree of repairs. That gluttonous wave of workers repairing bolts and checking beams. A mockery of upkeep, all that attention!" Wouldn't that be a kick?

149

Received an email to the new survivor35w@gmail.com email address!

> Margaret Anderson Kelliher here. . . . I just saw your letter and the pictures to MN/DOT. Very disturbing. I actually live right in front of 394 on the North side of the highway. I would really like to walk to the piers you took pictures of sometime with you. Let me know and we can set something up.

Hi Speaker Kelliher!

I would love to. Name a time/day and I'll be there, early eves are generally good. Note, VERY RECENTLY they marked the pier off with white paint on the ground, four big box shapes. I am seeing some negative comments posted on the *Strib*, and the media have made several errors in their reporting, which has been infuriating and scary. I have spent the evening contacting several outlets to correct the errors.

I am proud of the action I took, and I am disgusted by the politics. I have family and friends from all parties. I wrote to survivors and families to please think critically about what they hear. I told them that I "hope that you will view my letter and photos and compare that with what you know of me, before making any judgments."

Republican, Democrat, or Independent, we all need safe bridges and roads.

Despite my utmost attempts to pair with a nonpartisan THINK TANK (for god's sake) the BS is astounding.

However, Rachel tells me to focus on the people who have been levelheaded.

Would love to chat. Name a time, I'll come to you.

Kimberly

Kimberly,

What is your morning schedule like? I can check my availability for the end of this week and we can see if you can swing off on Penn we could probably walk to where the piers are.

Tell me about the errors in the reporting—since I will have some media contact tomorrow. I will check email in early a.m.—since I have been up for a long day here and need to get kids off to school in the early a.m.—I will check in with you tomorrow.

Thanks for your work—it is hard with the media at times and please don't read those posts on the *Strib*—and don't respond—these are the same few people who really don't seem to do anything but post on the blogs and comment spaces. Hang in there.

Is there anything you think I should say tomorrow with the opening of the bridge? I am not going to be allowed to speak by the Governor and who ever organized the event but I am sure I may have an opportunity to say something after to the media.

MAK

Hi,

You can use this if you like: "As MN opens a new 35W bridge, we need to remember there are many existing bridges in disrepair that need our attention." That was what I said at the top of the press conference. (Some media chose to ignore this.)

Or you could say, let's focus on the real issue—emphasizing that this shouldn't be a partisan issue. With MN/DOT reporting that they're $2 billion behind PER YEAR, for economic reasons as well as for moral reasons we must start catching up. We all need safe roads and bridges. (Side note: The closure and detours cost the economy an estimated $400,000 per day in added travel and other expenses. We spent $225 million to rebuild the new bridge. It would have cost $2 million to reinforce the plates.)

For you, Tues.-Fri. I will gladly clear time in the morning.

I should warn you—MN/DOT is on the defensive and we may get there to a locked gate. We might have to view the piers from the Dunwoody parking lot.

As for the media, MN2020 and I were very deliberate in focusing on the facts and my three examples—the "bricks & mortar" (the deterioration at hand).

I got on the phone tonight to the *Strib*:

1. The headline said I "showed off" the 394 pier. I said it sounds like I was standing there boasting. That is incorrect. Please delete the word *off*.

2. They said that the timing was "an intentional shot" at the governor. That is incorrect. I sent it to MN/DOT on 8/19, again a week later, then again (a 3rd time) a week later, then copying ten director-level people at MN/DOT. On 9/2 Susan Mulvihill replied that they received it and forwarded it to the appropriate department for a response. That is the last I heard from them.

3. They misquoted me. "Creating a memorial and having ceremonies are good, but we need to" and they said "move forward." I corrected them. I said, "go further." The former sounded like I was being ungrateful.

4. They said I was "refusing" to say where I saw the construction worker. I stated in the press conference (from the top) that the worker's request to protect his identity was what made me decide to document these myself.

And in general, I was holding my breath tonight. Channel 5 has put a whiney spin on my stories before. The 5 o'clock showing wasn't half-bad, but the 10 o'clock showing? It was all about me and my desire to "take on bridge safety as my life's work." WHAT? I didn't want this, it came to me and I explained that. Chris O'Connell also got at least one fact wrong. He said that all of the bridges that I highlighted are on the schedule for repair. Conrad deFiebre's piece notes specifically that they aren't on MN/DOT's lists. They just added this, or Chris extrapolated this info. The pier that I photographed wasn't marked with MN/DOT's flags last Wednesday (9/10) when someone from MN2020 was out there.

He said MN/DOT "had email problems," but this was never communicated to me.

Best,

Kimberly

P.S. I'm going to email you the transcript of my talking points so that you can see the whole of what I said at the press conference. It is important since no one airs all of it and will give you a better, fuller picture.

P.P.S. I should add, I got successful delivery receipts on each try.

Well, clearly they are defensive, and someone was not paying attention.

150

The new 35W Bridge opened on September 18, 2008. Survivors were offered the opportunity to walk across the bridge before it officially opened to traffic, but nothing about this felt right to me. This, again, is where each survivor is discretely unique in their responses.

I rail at the memory of the uplifted road that ended in midair, how that contrasts with the new high-tech, motion vibration, catastrophe-catching, pulse-taking sensors, of a bridge complete with five lanes instead of four and two elegant wave sculptures—each made of triple-striped concrete squiggles—at either end to mark where a passenger in a car might not realize they're crossing water.

I don't forget. Under a perfect moon glow and hue, I look upon the new bridge and compare it with my memories. Memories illuminated with carnage, the dramatic opposite of the sculpturally elegant and graceful lines, the scientific sturdiness and the low rail that shows us a clean shot of the river below. Never mind that it's December 10 and we had our first real snowstorm and it's only 7 degrees. Never mind that the wind cools a person's skin to what feels like minus 25 degrees. Could be summer, with a hot sun. All of this expensive concrete, so strong, so easy to purchase. At a price tag of $225 million plus a $25 million bonus to the contractor—112 times more expensive than the $2 million repair that could've saved us the terror. I turn right and head down the slight incline of the entrance ramp and see the short burst of progress that all of those cars speed over. I see all that pretty blue, and I shiver. I shall never get used to this monstrous sight. This sturdy, smooth, excellent bridge. I should've been ecstatic, in my city that had not taken care of its bridges. The fact that my city was spending money to construct a good bridge should have been seen as a step in the right direction. Logically, of course, it was. But for me it was hard to celebrate. It's complicated. I hadn't gotten over the nagging feeling that it should have never happened.

The new bridge strode upon the riverbanks, oozing verve and panache. His sinewy pectorals rippled as he swaggered toward me, chucking me under the chin. "Don't you worry, little girl. There there." I recoiled from his macho demand, hundreds of millions of dollars of raw materials purchased to assuage my fears. Via the new bridge, the masses will speed north and south. The public would no longer need to be bothered with collapse. It helped us forget. If only we had spent $250 million spread among all our other bridges. Elderly weightlifters. How they could benefit from an infusion of moxie. The new bridge leaned down to whisper in my ear, trying to soothe me like a medicated paste.

I was living in the past, in the hurt and the fear and the grief. The collapse lived on in my post-bridge body, in my heart, my mind, my skin, my wishes, my hopes, my fears. Ultimately, none of this mattered. My city was moving on without me.

151

When MN/DOT finally responded to my "Dear MN/DOT" letter, the sender apologized for the agency's nonresponse and claimed that it was experiencing "computer network problems last month resulting in the delay or loss of most emails sent to us from outside MnDOT." *Uh-oh. Red flag.* Then the sender expressed gladness that I had survived and appreciation that I "took the time to write to us to express your concern over the condition of bridges in our state." The letter writer continued by specifically addressing the examples I had cited, writing: "Since our bridges are exposed to the harsh weather elements in Minnesota, they will display deterioration over time. With preventive maintenance and routine repairs we can keep our bridges from reaching a condition that we consider 'poor' or structurally deficient for much of their usable life." *What I've been saying all along, yes! but common ground is encouraging.* The rest of the letter, however, expressed that for the Cayuga Bridge, guardrail, and pier, the deterioration I documented "only served as cover for the outer reinforcement and is not of structural value." For the Cayuga Bridge, the agency wrote, "The daylight you observed through the deck in the photo you submitted appears to be at an expansion joint where it is intended that there be an opening in the deck." *Wow, so there's supposed to be a hole in the bridge deck. Nuh-uh.* While they were the experts—the engineers and the certified bridge inspectors—the content and tone of the letter left me with a vague dissatisfaction. In each instance MN/DOT said the deterioration was noted and would be repaired at a time the agency deemed fit—as if I were being foolish to be concerned.

In my unscientific estimation MN/DOT does an excellent job of snow removal and treating our roads during inclement weather, such as during a dangerous 2016 Christmas winter ice storm. But the tone of the letter told me that people at the agency were viewing me like the poor little bridge survivor, trying to placate me. I didn't want to be placated. I wanted them to acknowledge that my concerns bore validity. I didn't need to dwell on missteps, but I wanted a pledge for improvements. I wanted them to talk about difficulties they faced so that we could see the truth, come together, and attempt to improve the system. The agency followed up the letter with an offer to meet with me in person. We had planned a meeting for October 18, 2008. In the end I canceled the meeting because it was clear my message wasn't getting through.

Also on October 18, I received an email from a reporter:

I'm Sea Stachura, a reporter colleague of Dan Gunderson's. He's mentioned that you've stayed in contact regarding bridge safety and the 35W bridge. A few months ago he provided me with the photos you took of bridges in disrepair around the state. He speaks highly of your commitment to honesty and safety.

I am the lead reporter on the NTSB investigation into the collapse, and as you likely know the board will release its conclusions on Nov. 13th. I'd like to talk with you about your thoughts and expectations leading up to this hearing. Would you be willing? I'm based out of Rochester, but I'd be happy to come to the Twin Cities and meet you.

After spending two hours with the reporter, my only contribution to the article was basically, Will we know the truth? In a nutshell I expressed my lack of confidence with the NTSB based on the misleading statements from the press conference on January 15 by Chairman Rosenker. I conveyed my hope that the report will be as objective as the NTSB is supposedly known for, but I expressed my dismay at the chairman's misleading statements that the investigators had "never seen this before" (too-thin gussets) when there is evidence to prove that they had (in the case of the Ohio I-90 bridge that nearly collapsed on May 24, 1996, reports were written and that bridge studied by ODOT and FHWA), and there were pictures of the bent gussets on I-35W in fall 2003. I also cited the claim by the NTSB that corrosion/maintenance had nothing to do with the collapse and said I hope it isn't a politically influenced decision that will support continued disinvestment in our infrastructure.

152

I decide to join Rachel on a Christmas visit to Texas, so I return to Barb and work on my fear of flying.

Barb says, "Let's start EMDR with a statement of how you feel about flying, something that represents the concern that we'll work on."

I offer variations on a theme: "Flying freaks me out."

In her therapy office, with its purple walls, I settle in with the Advanced

LapScan 4000. Sitting cross-legged on the loveseat, I don headphones and hand nuggets and watch the lemniscates. Some realizations: danger isn't just on planes. A person can be injured in a plethora of places in their lives. Or it could be internal, a disease that festers. My post-traumatic struggle with flight is due partly to sensations the body perceives: vibrations, dropping, and falling—all similar to what I experienced on the bridge. The position at takeoff is just like the position in the car when we landed on the broken bridge. We begin.

"I'm scared the plane will fall out of the sky." *Buzz buzz.*

"Close your eyes and say everything you notice with your body when you're on the plane." Lights: *zoom zoom.*

"I feel claustrophobic. There are people on both sides of me. Outside our row people chat, read, laugh, work on laptops. Babies cry, kids play, some fuss." *Beep beep.* Lights illuminate the ceiling. Flight attendants push a cart. The seat beneath me is rigid, lacking cushion. Magazines in the chairback in front of me. Hidden in the smooth panel above my head, oxygen masks would drop in the event of an emergency. In the overhead console, lighted symbols for seat belts, air, and a reading lamp. The shade on the airplane porthole, pulled down a third. Outside, grayish blue. We fly above the clouds into nothingness. No land visible. Vibration in the floor at takeoff has dulled to subtle shifting movements, slight variations from flat. I notice it in my feet. I've planted my sandals on the floor, but my mind knows the floor isn't real, not really. Beneath the floor isn't ground but ether. The engine's constant drone reassures, but I want to get off. I understand, even appreciate, that some people blossom in this environment. Here, where they relax, glad to relinquish responsibility for the duration of the flight. *Buzz buzz, zoom zoom, beep beep.*

Barb says: "You wouldn't want to fly that plane. Think of pilots and how you trust them to fly the plane."

"But I don't. I don't think I trust them. I don't trust anyone." I stop cold. "Wow. That was deep. Where'd that come from?" Then I answer my own question. "Probably from childhood."

Barb nods, acknowledges that I've done much "Mom work" before in therapy.

"When there's turbulence or when we drop, I rattle super fast, 'I'm okay I'm okay I'm okay.'"

Barb says: "That would be nice, if it worked, but it takes a Herculean effort for the mind to overpower the body. What if we tried to find a new statement you can tell yourself? How about 'I can handle this'?"

Strategies when on the plane: Be present. Do not think about the future. Do not think about the past. Not mind-boggling thoughts by themselves but with the sensory stimuli of EMDR, this message does more work. The corpus callosum throbs with possibility. Tears well at the corners of my eyes, as I feel myself begin the impossible process of letting go.

Barb offers more ideas. "Have distractions. Talk about all the fun you'll have, the places you'll go. New phrase: 'I can handle this.' When the body reacts, try to do the opposite."

I say: "Part of me just wants to be okay, to not even need those drugs. Just get on the plane, cognizant and alert and well."

"It's a choice," she says, pausing. "Know when the sensations come, they're not related to the bridge. No doom."

My mind formulates futuristic solutions. It dreams of a car that can fly when the ground becomes unstable. In the meantime EMDR drops lifelines to my strained grasp as I cling to the ground. I continue working through my flying fear, which I realize is more a fear of crashing.

"I'm afraid of being in pain," I say.

"That ambivalence comes from negative beliefs. Maybe it's time to let go of the old story."

I can't even comprehend this. I repeat the phrase as I try to take it in, feeling my shoulders drop as I imagine being reborn: innocent, bold, positive.

"I wish I could be more fearless, more positive. I want to be fearless. To not think, when things go well, that something bad will happen. I know that's the goal." I imagine the bumps in the sky and the plane staying up. Rachel sits by me, and we talk. I feel turbulence, but instead of my hands tensing and my stomach rising and falling, it's actually pleasant. Everyone acts normal. But are they clueless? Is their calm evidence of naïveté? We can crash and all die.

Reality check: imagine turbulence again, like a roller coaster.

I come up with statements that stun me. I'm afraid the plane will drop out of the sky. BUT IT PROBABLY WON'T. The only thing I can control is my worrying. When I feel fear, I know ways to relax.

Barb asks, "What does your inner wisdom say?"

"It says, Get on the damn plane and go. I rarely have nightmares about the bridge now. If it were only as simple as encouraging myself . . . could I know that I'm ready for this? No matter how it turns out?"

"You can change your mind or thoughts and feel your emotions, but the body takes longer to release the trauma."

I think about this and realize how much it helps to understand this mind-body connection, that the process is slower than what my mind commands and that it's normal for the body to feel upset afterward. I get it—relief isn't instant. It takes all of the senses. It takes a brain bridge. Though it isn't easy to go to these places, to face situations head-on, Eye Movement Desensitization and Reprocessing is a gift. My progress surprises me.

Barb continues: "Often our biggest fear is the unknown, so if we can even imagine the unknown, it feels less scary. When you can imagine your biggest fear, it loses some of its power."

This means I have to imagine my own death and funeral. I ask, "Imagine despair?" Yes, Barb says. She says pick five people you want there, plus a celebrity. Imagine what they'd say. Another survivor described once how she had relaxed through the fall. I figure out that the way to bridge my fear is to touch my death. Be okay with the consequences. Let go. And fall.

153

Partial Transcript from One of My Actual Sessions of Eye Movement Desensitization and Reprocessing

Therapist: Barbara Stamp, LMFT
This session's work: (1) Fear of Flying; (2) Crossing the Rebuilt 35W Bridge
Date: October 18, 2008

Objective: To demonstrate the practice and work of EMDR

[Me:] One thing I was thinking about last week was, It would be so nice if I could just get on a plane and not be scared. Or go on a vacation and go on a cruise and be on a boat and have it moving and not be scared.

And we've come close a couple of times to actually considering doing that, and it's just way too big of a step.

I don't know how to get there. And even when we just go out . . . like out of town to a cabin or something. Like, I just sometimes feel so . . . um . . . it's hard to explain. I feel restless and uncomfortable and uneasy.

[Barb:] When you're away from home?

Yeah, it's like, I just . . . don't want to be there.

All right, let's work on those fears. Notice any change on elevators?

Maybe a little?

Okay, and it is a process . . . but a little is good. Because very often, when somebody has a series of fears, you work on one—a little bit better. Work on another one—a little bit better. And pretty soon they all kind of click in. And you're fine with all of them.

That would be nice.

Yeah, it's weird. Because fear is at the basis of all of them.

[EMDR starts on the Advanced LapScan 4000.]

[Green dotted lights move against a black background in an Infiniti / figure eight pattern.]

[Alternating pulses vibrate in both palms.]

[Alternating buzzing sounds ring in each ear.]

Just feel in your body getting on an airplane.

I feel dread, claustrophobic. The floor feels hard, tinny, and the aisle is so narrow, and all these people are staring at you as you walk down the aisle. And it's just, ugh, I just hate it. I hate it.

I'd be saying to myself, Oh god please make this be over with fast. Please let us get there safe.

I would get into my seat. Get situated. Everything's fine while we're on the ground. It's once we start moving . . . I mean it's still full of dread.

Plane takes off—feel it in your body.

My stomach feels tight. And I just notice every movement. I'm just really tense the whole time. And I'm just really scared the whole time.

So feel that in your body and amp it up. Get into it. This is about desensitizing. Breathe into it.

I feel sick.

Breathe through it.

I don't know what to do. I just don't want to be there.

Feel all of it in your body.

I'm not ascending anymore.

So you're up in the air. How does that feel?

It feels so weird, like we're just hanging out in midair, and my mind wants to relax.

Okay, let your body relax.

But I think I just overthink. I think, How can this be right? How can I be up here? How can this be working? It's not going to work.

Pretty amazing, isn't it?

Yeah.

Well, it's been working for decades, so it will probably continue to work. [Laugh] *The statistics that you're safer in an airplane than a car. So just feel that. You're up in the air, and you're just hanging out. It's just amazing what goes through your head. Ground is way down there. It's like you're just floating along. Feel that in your body.*

Okay, as long as there's no turbulence and as long as I'm distracted . . . the whole time. As soon as there's a little bit of turbulence, I start tensing up.

Feel a little bit of turbulence. Breathe into that.

I'm gripping the armrests; my stomach gets tight again.

Let your rational mind think about how many airplanes all over the world are in the air and get down safely. What a vast world we have.

I flash from a little turbulence to suddenly plunging down.

Feel that, Kimberly.

It's horrible. I visualize the scene from the movie *Fearless* with Jeff Bridges where the plane was going down and everything was flying everywhere. Can we stop?

Let's get out of turbulence. Back to smooth.

Nervous but okay.
Breathe through it knowing that you'll be on the ground. Feel how that feels.
Relief.
Thank god.
Going to the curb where Rach will pick me up. I feel really relieved, but I feel really crabby that I had to go through that. Just riding in the car.
[EMDR with the Advanced LapScan 4000 concludes.]
[Session continues.]
So, you did it.
Mmm-hmm. That was not fun.
It would be normal to feel vulnerable—you're not the pilot. But allow yourself to feel that, so you're basically desensitizing yourself because most of the time, I believe, it goes pretty well and sometimes just amazingly smooth. I mean, not very often do we hear about something bad happening.
It doesn't matter to me.
Pardon?
It doesn't matter to me, though.
What?
It doesn't reassure me that bad things don't happen that much.
Right, your rational mind. It doesn't help your rational mind?
It doesn't because bridges don't usually fall down. So to me I don't trust it. And I don't trust other people's rational reasoning. I feel like my mind can create catastrophes anywhere I am. I can be downtown, you know, walking through the skyway and imagine a catastrophe. I can be driving to work, and I can imagine another one.
That's very common when you've been through one to imagine another one is going to happen. Where? When? Okay, so how are you doing with that?
I'm stressed out.
Feel the stress. Let it talk to you. What does it say?
I don't know. I just feel kind of overwhelmed.
Ask the overwhelmed, what does that mean?
I'm not getting anything . . .
Being in your safe place . . . can you switch gears?
I don't know if this is my safe place, but I like cooking.

Make it very real so you can recall it. In your kitchen cooking. Smell it, taste it. See it. Hear it . . .

A wok on the stove, I have olive oil in there, and then I can smash up some garlic and chop that and then put that in the skillet to infuse. And chop up some veggies to stir-fry, salmon in the oven. I like doing stuff like that . . . something very comforting about it for some reason.

Sounds very nourishing.

Yeah . . .

This would be a good image to hold. Yeah? So, when you are . . . wherever you are, you can just imagine changing channels. Obviously, if you're in a place where you have to do something about something scary, that's different. But if something is coming and you're looking around and you're thinking, Wait a minute now, I really have no reason to think this, try to change channels, to one of your safe places.

Like if I'm flying and there's turbulence, try to think about cooking?

Yeah. Because what we're thinking is very powerful. I mean you're up in the air, and all of a sudden you feel a little bump. Now, you can either go into fear and say, Oh my god, we're going to crash, and then you picture the crash and all that stuff. No. You don't know that. So, no. [Tell yourself] I'm going to imagine being in my kitchen and cooking.

I used to really resist that because I thought . . . I mean it's kind of ridiculous because it's not like my thinking—either way—will change what the outcome is?

It will change how you feel.

Yeah . . .

I mean a little bump—we really don't have to panic.

But, I do. And then I think I have to be vigilant and nervous and tense and think the worst so that when the worst doesn't happen, I can be lucky.

No. No. That's too much work. I mean, if you absolutely know. I mean, it's possible something could happen—when you're in the air and there's a little bump. We can choose to not go to the disaster. We can choose, "I'm going to be in my kitchen cooking." And that's true for anything, whatever we're doing. And some little surprise happens. Walking down the street, twenty kids come, and you can think, Look, they've got four adults with them, and they've got this little rope. So they're probably totally safe. So I can focus on that. So it's usually our reaction to things that can cause a problem. And there's not the evidence that something's going to happen.

I'm too much of a realist...

Realists are often very cognitive and hopefully RATIONAL. *Right?* [Laugh] *Because our minds... can just... go all over the place. If we can basically control our thoughts: Wait a minute now, there's absolutely no evidence that this will happen, so I choose to not go down the negative path. After you've been in some kind of crisis, it's hard to do that, but you can train your brain.*

It's harder for you, perhaps, because of how you grew up, with Mom, never knowing what in the world was going to happen to her. So I see the tendencies from childhood carrying over into adulthood. That can be changed.

And it's not about easy; it's about shifting gears and saying I have choice here. Then you don't feel like a victim.

There's a part of me that wishes I could just get on a plane and get it over with. And there's another part of me that wants to stay far away from it and never go there. Just like with the new bridge. Part of me that just never wants to go over it.

They both feel real. They're both true. So just imagine holding them both.

Rach and I talked about driving over the new bridge. I said that it was like a graveyard [Tears] where people died. I don't want to go over it. Then one day I had a change of heart and felt, I just want to drive over it and get it over with!!!! It'll just take, like, thirty seconds!!! We can be over it, and not think about it anymore! And I asked Rach, "Have you gone over it?" And she said, "No, I haven't gone over it yet." And then she said, "I want to go with you." She wants to be there with me when I go over it for the first time.

Right. That would be good.

Because she wasn't able to be there before.

So that would be very good for both of you—because, in essence, you have a different memory then.

It makes me want to cry just thinking about it. I didn't feel anything before, but now that some time has passed...

It's coming up so that you can heal it.

Part of me this whole time didn't want to go over it because I didn't want to support the fact that they built this new bridge, too little too late. I wanted to symbolically show my disapproval, for me. I knew it wouldn't mean anything to anybody, except me. I told Rachel that I didn't want to go until it feels right to me. And it hasn't felt right.

I would support that. It's got to feel right. You've got to be ready.

But now it's kind of like talking about the airplanes and the elevators and everything else ... maybe if I got it over with, it would be in the past then. We could let it go.

Just make sure you're ready and make sure she's with you. It would be healing. The two of you would have a different memory.

[I'm staring more intently now at the EMDR board.]

"I'm on this bridge, but here I am with Rachel. Hold that image." Now hold the image next to it of "I'm not going to go on the bridge." Hold them both at the same time—like they're on two plates. See what it feels like—which one is heavier, lighter?

I want to go over it.

See what that feels like?

That's weird!

I think you're getting ready, Kimberly. Feel them both—which one has more energy, more light, more whatever?

Rach and I like going to dinner together because it's nice, relaxing, bonding. We like going to new places. We also like repeating the same things that we like. Erté—feeling what it's like to just be alive together. We should do it tonight. Because we were trying to figure out what to do tonight.

Hold that in your body ...

I feel like I won't be as held back.

Part of your healing, Kimberly, is to create a new memory. Notice your feelings.

It feels more powerful for me now without any media, cameras, and any of that. Only thing I'm not sure about still is I kept saying I didn't want to go over it. And I really didn't because I felt like they built that. You know, too little too late, and I don't want to go over it.

So think about that. They're not going to know, and they probably don't give a rip. The important thing is, What will serve you?

I see myself going over it and being mad. Looking at how pretty everything is, how perfect this new bridge is.

That's all right. Just honor your feelings—you probably have MANY of them, many different thoughts. Both of you would probably have a variety of feelings and thoughts.

I pictured driving up to it and going over it . . . "Now, I'm kind of tired!" [Laugh] Emotional stuff is tiring.

The whole time you're reprocessing.

What's weird is, Why isn't it enough just to think about it? Why don't I make progress? I mean, I've thought about this stuff ad infinitum, and why would just coming in here and talking about it for an hour change it? How does that work? I don't understand.

Because you're ready.

Is it because I'm closing my eyes?

I think that when you do internal work, you get your OWN information that's separate from up here. [Barb points to her head.]

"You mean it's coming from my heart more?"

Well, it's coming from every cell in your body, I think. Mind, feelings, body, soul, spirit—when you're in imagery, that all, I think, connects. You're in that alpha state, so unconscious stuff can come up. Most of the time we get something stuck in our left brain, and there it sits.

In our logical heads?

Absolutely. And there it sits. When you close your eyes and just allow everything to come: all the feelings, the sensations in the body, everything we've experienced in the body. So, when you do this, you are releasing those feelings in the body.

154

After therapy Rach and I decided to get it over with. So I drove with Rach toward the new 35W Bridge. The freeway got curvier as we neared the university.

Before this we discussed what song to play, so I picked "Bubbly" by Colbie Caillat because it had soothed me right after the collapse. In the months that followed, I used to hear it on the radio frequently. What would happen? I wasn't sure, but then this song would come on the radio, and for a second I'd be transported to a time when I felt safe, a time when I didn't think about bridges, a time when I could take that safety for granted.

The song would go, "'Cuz you make me smile / Even just for awhile," and the lines resonated as I'd think about love and loving someone no matter where life takes them. This song always made me think about Rachel and how tenderly she'd hold me to soothe me. How lucky I am to have that kind of love in my life. Every day, whether I was just coming home—she and Lucy running to greet me—or at night before we'd go to sleep, she was the one constant in my life, that place of home, where I didn't have to long for the safety I'd lost. Nights when Rach had to study late and I'd go to sleep before her, I'd sing only half-joking, "Tuck me in! Tuck me in!" And sweetly, she'd indulge me, saying, "Yes, baby."

Driving toward the bridge, I felt quiet thinking about these moments and reflecting on the time I've had since the collapse, thinking about people who've passed—my mom, Rach's grandma—and wondering about the future.

"We're getting closer," I commented, then in what seemed like the next breath, we started over it—the perfect, intact, wide expanse of concrete, the car almost gliding as Colbie sang about being tucked in "just like a child."

The bridge must have seemed longer in my mind because before we knew it, we were suddenly done crossing it. Maybe I thought it was longer because for so long I wasn't ready to cross it and never planned to, or maybe it was from watching that first crossing on TV—when I sat at the kitchen table, tears rolling down my face, or maybe it's just the eerie knowledge that those early-morning drivers (people featured on the morning news shows who were first to ceremoniously drive over) were rolling over a place where so many had suffered or died.

Whatever it was, we were done, and I was surprised. With wet eyes and pretend disappointment, I said, "Well that was boring." We both laughed, feeling the tension wick off of our shoulders.

I said: "We did it! We did it together," a moment later adding, "Let's go over it again."

Rach gave a cheerful "Okay."

So we exited at University Avenue, took the streets around the block, and then headed back down the on-ramp onto the new bridge. On the return crossing I looked out my window and saw the sky that had turned a deeper orange, set against the deep and dark blue of the sky.

Quietly admiring it, I whispered, "Look at the sky."

Rach whispered back, "Pretty... how do you feel?"

"Good," I said as we headed toward Tum Rup Thai, a recent favorite, where we had planned to celebrate. "Weird... but good."

155

The day before Thanksgiving, I had an appointment as part of the 35W Bridge Fund settlement process with Wil and Special Master Steven Kirsch, who was one of three judges who had until February 28 to review 179 cases and determine how to divvy up the compensation fund. The interview to discuss my claim took place in a downtown St. Paul high-rise, in a conference room, again with a wall of windows, this time overlooking the Mississippi River. With recognition of how many times I'd told the story, Mr. Kirsch asked if I could say in my own words what had happened that day. Back to square one. Telling had become rote so I was able to handle it with poise.

From a finite pot the legislature tasked the panel with divvying up $38 million. Split 179 ways, it certainly wasn't a windfall. But it was help, beyond the short-term, for consequences caused by the state's failure to protect the public, help for those who could no longer work as a result of injuries, help for spouses who had lost household income—many incomes cut in half—help with bills for ongoing therapies, medical care, and surgeries.

I met with Special Master Kirsch. It was fine. It was easy in comparison to the hearings, testifying, and media blitz I had been through.

I felt strange talking about myself so much, but it was also vindicating in a sense to finally acknowledge the devastating effect the collapse had had on *my* life.

My lawyer told me that the atmosphere could be as relaxed or formal as I preferred. I chose relaxed. Our lawyers (bless them!) had done most of the work for us.

It was comforting to have Rach with me, to acknowledge this and go through it together—after all, it affected her too. Completing the compensation process was a milestone that we were relieved to put behind us. Afterward we went out to lunch to celebrate making it through another step.

156

At the legislature, during the debate on the House floor, GOP representative Mary Liz Holberg pushed for public disclosure of all "awards." That wording always struck me as asinine. As if being terrorized, injured, and then forced to fight for help were so rewarding. My pro bono lawyer, Wil Fluegel, kept me updated on the progress of my claim and explained that the actual records wouldn't be released, but there was a threat that because it was taxpayer money, the public had a right to know who settled and at what amounts. Debate ensued on both sides and was quibbled over in the House chamber. These people did not ask for their bridge to collapse. How much invasion of privacy was necessary? Those who tried to protect survivor and victim privacy were forced to compromise. Thanks to Representative Holberg, it was written into law that each disbursement would be publicly disclosed: the claimant's full name and the amount of compensation.

157

After the collapse, let's just say I was tense. Trauma doesn't only create new problems; it brings old problems to the fore. Anytime anything physical happened with my body, I panicked. The 35W Bridge collapse *lived on* in my body as gradually, over many months, I realized that I uncomfortably, for no reason at all, tensed my... you know... the thing that the comedian Sacha Baron Cohen, of *Borat* fame, called his "ah-noose."

Mine's tense. There, I said it.

Strange, personal, secret—at least until this writing: while I unconsciously experienced the stress of trauma in this sensitive and private place, I willed it to relax. When that was only mildly and briefly effective, I tried different tactics, silently annoyed, talking to it as if it were a thing, not a part, saying things one might whisper in a classroom, in a voice so low that almost no one hears.

"Stop it," I pleaded. "Relax."

I walked around for weeks thinking, My ass is tense. And I couldn't tell anybody because it was embarrassing as all get-out. Right? Why would you tell anybody? They'd give you so much crap. Under high pressure the pain felt like I was trying to *make diamonds* down there.

With friends, or at work, I can tell people that I clench my jaw at night, that I'm getting a mouth guard because I'm hurting my teeth. I can tell them my molars hurt and this will worsen over time if I don't do something about it. I can try to explain the PTSD. Tense. Scared. Vigilant. "Makes sense somewhat," people say without words, with blank nods.

I know about having PTSD but less how to fix it. The unconscious—the body—behaves as if it had a mind all its own.

My actual mind, my brain—a pathetic, out-of-shape, bench-warming player in this whole drama—receives commands. I send messages to my muscles, and for a short spit of time, the muscles obey—only to succumb a short time later to the body. I'd felt a persistent nonspecific throbbing, back there. One night I lay on the couch and grimaced, my tailbone or some vague area was tender like I'd bruised it. But I hadn't fallen or been hit by anything. The couch was soft. Now I was freaking out. What if something was really wrong?

Pulling into the surface lot one day at work, I turned the car off, unbuckled my seat belt, and gazed at the loading dock, and then I debated. I felt the weight of my cell phone in my palm and thought, Make the call? Maybe it would get better by itself. Or maybe it was dangerous to ignore it. I had to do something, but was it worth the embarrassment? How would I describe it to some stranger on the phone? A long holiday weekend was approaching, and the pain was worsening. Post-collapse, I tried to compare uncomfortable or scary things to falling with that bridge, so most things paled. I asked myself, what's worse than death? Is asking for help worse?

I dialed the number for Park Nicollet Clinic. A woman answered. "What can I do for you?"

"Um . . . ," I stammered, "I don't know what department to call. I don't know what's wrong with me. I don't even know if I have a problem, but I was thinking I should see someone about it."

"Okay. What?" she said.

"I'm having this vague pain." Oh god, here I go. "In my rectal area, I guess." I closed my eyes and held my breath, waiting for the world to end.

"I don't know what department that is. Proctology, I guess? Can you hold?"

"Okay, but before you go, this wasn't easy to explain to a stranger, so can you please make sure that I don't have to start all over with someone else?"

At the actual appointment they'd sent me to a GP, and I had to explain my symptoms again to a triage nurse, who dutifully jotted notes on a clipboard, then inputted them into the computer.

"So you're having rectal pain? Where exactly?"

"It's vague," I said.

"You can change into the gown on the table. The gown opens in back. Take everything off from the waist down, and the doctor will be in shortly." When I transitioned from my clothes to the hospital gown, I felt a wave of cold air. Getting on the table, the tissue crinkled obtrusively like it relished giving me away. I imagined, if the tissue could speak, it would announce to everyone in the immediate vicinity, outside the door and surrounding the room, Hey, the girl in here has an ass problem.

A knock on the door.

"Come in."

"Hello," the doctor said. Based on her nametag and accent, she was from Russia.

"Vare iz yer pain from? Vare are you feeling za pain?"

I answered, thinking, this is the umpteenth time I've talked about my backside.

The doctor said, "I'll be right back." She scooted on the wheeled stool to the counter, where she donned medical gloves and retrieved items that I couldn't see from where I was sitting on the table. When she rolled back, she said, "I have zis," and with gloved hands, she held up a clear acrylic thing that looked like a funnel. Through the neck, a matching stick with a rounded tip, excellent for gentle probing. As she presented these items for my approval or general knowledge, the thought crossed my mind that what was about to happen is what people imagine when they hear the words *cavity search*.

We were ready. Gloves, funnel, probing stick, K-Y. The doctor continued: I'm going to do zis, then I'm going to do zat, then, "I'm going to separate your cheeks."

I followed the rest of her instructions, turning on my side. The doctor proceeded like she said she would. Humiliated, I had the sinking feeling that I should've acted on my flight instinct and fled by now. Idiot, I should've waited. Maybe it would've gone away. She pushed it in, got a good look around, perhaps with binoculars. I don't know.

As I stared at the beige wall, I listened to her report.

"Ah. Every sing iz very healthy. I don't see any problems down here." She paused. "I'm going to remove the . . ." I could sit up, she told me.

"Thank you," I said when she let me up. And then I couldn't believe I had thanked her. But what on earth else was I going to say?

Then she said I could put on my clothes. After she left, I stood. It felt like someone had spit down my crack. I was all gooey. So much slippery lube. I wobbled to the tiny sink and, hand over fist, pulled tissues from the box and then mopped my backside.

When she returned, she said she couldn't see anything wrong so she felt there wasn't a need to order further tests at this time, but we could start a medicine and see if it helps. I went to the pharmacy and handed them the slip of paper. After I paid, I read the prescription. I couldn't wait to display this proudly on my bathroom counter! What a charming name for a medicine, a clever amalgamation of the words *anus* and *solution*. It came in a white tube with ANUSOL in bold, cheerful orange lettering. The instructions did not explain how one would administer said ointment by oneself, but I imagined, by a circus trick and a contortionist's backbend, and pluck and moxie, one could follow the dosing instructions—apply rectally three times per day—without issue.

But it wasn't getting any better, and it wasn't going away. The longer it continued, the more I worried.

My post-collapse paranoia resurfaces. Is something wrong with me? Cancer? What tests would doctors order to rule things out? My mind swims the gamut of possibility.

Later, at therapy with Barb Stamp, I wrestle with how to start. People go to therapy to work through their histories, circumstances that prevent us from living a full and self-actualized life—the things that have hurt us. So I decide to talk about it.

"I have this pain." I pause, search for words, procrastinating. "In my body."

"So tell me about it."

"It's really hard to. I can't." I try to tell Barb about a taboo part of my body that hurts. But my face flushes, just thinking about talking about it. Yet I'm torn because I'll worry and obsess, and that's not good for me either. I tell Barb how, with Rachel, I can say all kinds of crude things, but I can't tell you what this is.

After hemming and hawing—because I figure it'll be better than spitting it out—I retrieve a shred of courage. "Well, it's like I . . . my . . ." Okay, I tell myself. Deep breath. I press my lips together, think, Take a chance, just say it, it won't kill you. You survived a bridge collapse.

"My butt is tense."

"Ohhhh . . . ," she says, and instantly I feel this is a mistake.

"It's in my—" I forget who says something first.

Barb asks, "The pain's in your anus?"

"No, not there!" I blush. She asks more questions, and I clarify. Dread burns through me. But she isn't fazed. "Your body is trying to tell you something," she starts. And in all earnestness she says, "I want you to tune in to your rectum."

I burst out laughing, and she laughs too. We get it out of our systems, and then she tells me to feel the pain.

"It's dull and vague."

"So, if that pain could speak to you, what does it want to tell you?"

"I don't know," I say, my face hot.

She says in her uncanny and reassuring way: "Just listen to it. The mind is in every cell in your body. What does it want to say to you?"

This reaches me. I describe myself as if I'm talking about someone else.

"No one can see it. You'll ignore it. You don't want to acknowledge it. People don't see anything wrong with you. You look the same. But inside you're not. People are so cynical. They give you looks that say, What do you have to heal from?"

Barb says my pain is a metaphor for what's going on with me. No one sees it, but it's there. I had been to the doctor, so I know that nothing is medically wrong with that part of my body. She encourages me to just let the pain be there.

Earlier that day media reporters called to ask if I'd interview with them. "Are you accepting a settlement from the state?" they would've wanted to ask. Then the inevitable questions about how I was doing would follow. Physical, visible injuries, people understand. But my injuries were hidden. The dangerous media: I'd been steamrolled before, beat up by the thirty-second spot, by the summary and the headline. Back when we were pushing hard for the fund, frustrated by some interview experiences, I learned that reporters can take a story any direction they see fit. This sounds innocuous, but it amounted to spin. During one interview I had made it clear that my focus was the survivors and the Bridge Fund. When 99 percent of what I had said about the fund had been cut and the gist of the interview was "Kimberly goes to the chiropractor X times a week," I had to play damage control with other survivors.

Back in 2007, when I had the pleasure and privilege of talking with Cathy Wurzer on the set of public television's *Almanac*, I had described the survivors' injuries—my first foray into advocacy. At this point how could I convey my injuries in my neck, spine, and knee—my loss of height, my hives, my pain? Invisible. The anxiety, grief, short temper, poor memory, lack of patience, inability to concentrate, obsession with mortality, the post-traumatic stress. I warred to make peace with my body and my mind.

But it was all invisible.

Cathy was one of the reporters who called, asking if I would talk to her. How was my recovery going? "Well, Cathy," I could've said, "when things happen in my body, I have morbid thoughts. My emotions manifest in my body, and just this morning I got a rectal exam . . . And how was your weekend?"

Unlike when a bridge survivors' fund hung in the balance and every interview was a fight for survivors and those who died, I didn't have to do interviews anymore. It was time to tend to what I'd largely ignored during seven months of lobbying the Minnesota Legislature: recovering my body.

A news bite popped into my head. Someone like *Entertainment Tonight*'s Mary Hart saying in a saccharine tone, "Bridge Survivor's Poor Sphincter. Next on ET." This is how trauma manifests itself in the body. The pain was real. As real as you are reading this. Over the coming days I administered the medicine and continued with a trifecta of therapies—chiropractic, EMDR, massage—and this knowledge that I, in fact, was not dying from some kind of disease eased my secret pain. And over time that pain changed to a psychological happenstance that lessened with conscious awareness. Wellness would come, one sensation at a time, through patient acceptance, regular doses of humor, and a willingness to be human and gentle with myself and the strange workings of my body trying to recover from the worst crisis it had ever been through.

158

Walking in gently warming April air, leaves sing beneath us, my feet and Lucy's paws. We watch people jog with iPods or walk with dogs. We walk near the edge of Lake Nokomis, dry browned brush crunches under Lucy's galloping. She stops to christen a clump of dead leaves from last fall, past pine trees—their branches hanging in lines. Later I spy a kite twisting erratically from a high tree branch and think, "Another spring. I'm living."

Along neatly trimmed sidewalks and sturdy benches facing sunsets, I survive to see the unfolding of another glorious season of color.

Will every season be this way? Forever, with me always reflecting? Part of me hopes so, and part of me doesn't. And yet, like my body, my mind's not in charge. If I continue to be lucky, if I continue to have a life force, a reason for earthly pursuits, I'll come upon another fall, another winter, and so on, and have more stories to tell—less embarrassing ones, hopefully—of surrendering to my body.

Like last winter, when I shouldered a pain in my tailbone—a nonspecific aching that didn't go away in one day, in two, or even a week. I worried. This hurts. Why? Natural rhythms? A hemorrhoid? Yuck. Maybe something much worse. Colon cancer? My mind conjured medical diseases. One thought led to another, which in the end led to contemplating my own death.

Always. I go there, contemplating my own death. When will I stop this? Will I ever? Someday, guaranteed, there will be a day when I can't keep my body in balance, unable to fix whatever ails me. Someday, if I'm hungry, I won't be able to eat. If I'm thirsty, I won't be able to drink. In pain, won't be able to relieve it. What will that be like? I dread it. How awful will it be?

Will my mind be able to transcend the body? So far my mind is getting whooped. Score? Mind, 0. Body, 100. I look the same on the outside, but inside my body carefully maintains vigil. When walking Lucy, I scan the ground for holes and irregularities in the pavement. I must avoid falling or tripping, which would send a jolt like electricity up my spine. My body refuses to bend to my will. My mind worries.

159

May 26, 2009, I received a letter regarding the possibility of participating in additional lawsuits—this time against URS and PCI. Besides dreading opening myself up to more litigation and submitting to the opposition's doctors, who would tell me I'm not injured, the evidence has demonstrated to me that the State of Minnesota and MN/DOT were most at fault and didn't follow the advice given by URS.

As reported by MPR news: "Attorneys administering the state settlement found the victims had suffered more than $99 million in injuries and other losses, but they had only $39 million in public funds to work with." The outcome of these secondary lawsuits was a settlement in the amount of $52.4 million. URS of San Francisco agreed to settle to "avoid prolonged litigation"; the company did not admit liability or fault.

An excerpt from a report on the website of Minnesota Public Radio, an NPR member station:

> Minneapolis—An engineering firm hired to evaluate the Interstate 35W bridge before it collapsed in 2007 has agreed to pay $52.4 million to more than 130 people affected by the tragedy, attorneys announced Monday.
>
> The settlement with URS Corp. was the final lawsuit the bridge victims and their families had pending. They will take away $48.6 million from the URS settlement, bringing the total received from contractors and the state to more than $95 million.
>
> The state has paid the victims $36 million through special legislation. A $10 million settlement was previously reached with PCI Corporation, the construction firm doing work on the bridge when it fell.

I declined to participate in the secondary lawsuits against URS and PCI. The biggest reason was because I believed the state was the biggest villain in this whole mess, and I had certainly waged that war with my full being. My personal opinion remained unchanged.

As part of my decision to forgo participation in the secondary lawsuits, I did not sign a non-disparagement agreement. The implication was that I could speak and write, free from impediment. I could tell this story and identify the players by name, whether they be the State of Minnesota, URS, PCI, or others. This is an unanticipated result but one that pleases me, for principle and action in my case are as valuable as monetary compensation.

160

The door dings my arrival, and the dentist office smell smacks me in the face. Hate that smell. Dentist chemicals, strange and almost dry, unnatural, these chemicals have scientific names with plenty of syllables. Acrylic monomer, a plastic-like material in dentures and retainers; formocresol, a preservative; and there are others.

Patients wait in a cozy front room rooted by a fireplace. I sit on a green velour couch and peruse the reading materials and indulge in a *People* magazine. There's a cover article about a supermodel who, and I literally quote, "let herself go," and now she's finally come to her senses and has lost the weight. Stories about celebrities can make a person sick. So much superficiality and judgment, what a waste of time. Page through book reviews. Oh, Marlee Matlin and Cloris Leachman both have new memoirs.

A woman calls, "Kimberly Brown?" and I set the magazine down and walk down the hall. In a great building this dentist office reinvented itself from an old home on St. Paul's Grand Avenue. In the patient rooms big windows overlook pretty foliage outside and a mobile of leaves anchored on the ceiling spins slowly, with light from a recessed fixture coming through leaves.

"Are there any changes with your medical history since last time?"

"Well, no, but lots of changes with my teeth." Describing the clenching and the incredible sensitivity, I spend more time here lately than at the chiropractor. Linda, my hygienist, makes all sorts of notes and gives kind nods.

"Which teeth are having trouble?" she asks.

"Feels like all of them scream at cold and heat."

As the cleaning progresses, mostly the molars on both sides, top and bottom, scream most. Linda sprays water to rinse, and pain shoots to my gums. On the last lower right molar, a crack extends down the enamel. This is bad news. When I had a molar capped years ago, I gripped the armrests until I thought I would rip them off. To avoid dental pain, I brush and floss almost obsessively, but this cap is only a few years old, and it's already having trouble. Linda scrapes and polishes, and the leaves spin. With the constant application of pressure at night, she confirms, the structure of my teeth are taking a slow beating.

"It's no guarantee, but try the night guard for a while and see if the sensitivity settles down."

My teeth problems remind me of bridge terminology. My teeth like Minnesota's bridges, structurally deficient, stress fractures, stress on bridges caused by increased loads and capacity. Except, without knowing it, I put pressure on them at night. Can they withstand the pressure? Clenching day after day, they show signs of stress. My body warns me. Will I listen?

When I do this at night, unconsciously, can I change the trajectory of what's bound to happen? Like bridges, we start with diagnosis, or "inspection." But as master of my own health, "inspection only" is no option, the way MN/DOT maintained the 35W Bridge. Just because problems aren't visible on the surface doesn't mean they aren't there. This is lucky knowledge! Knowledge I can use to right my life, to keep it on track. I don't have to give up and resign to pain, helpless against it. I take steps. Mouth guard, step 1. Clenching at night, no longer enamel on enamel but enamel on rubber (actually ethylene vinyl acetate, EVA, or polyurethane, PU).

I'm paying attention. I need to keep working on my subconscious, the anxiety. It's a mysterious, slow process, but ignoring it just guarantees destruction. My dentist eventually would put plaster in my mouth to form my very own mouth guard, which I would wear nightly, indefinitely. Six years later, in 2013, there will be a hole in the back molar, but better there, in the mouth guard, than in my permanent teeth.

Repair now.

My family, me, survivors, we all can learn. We don't have to be ashamed. Life throws curves, inevitably. How we deal with it, that's where we show our mettle.

161

In the terminal, announcements, moving walkways, escalators; conveyor belts carry bags. Wheels on luggage click as they're pulled. Passengers in motion. Red laser light as the machine reads the data. *Beep.* "It's getting closer," I hear it say. The line advances. TVs overhead broadcast CNN, a derby somewhere. My turn. *Beep.* The ticket machine says, "Good luck. You can do it." I smell shampoo and cologne. Foreign smells show me I'm not at home. Scents mix like flavors in a jar of jellybeans. Slick surfaces, industrial low-pile carpet. Walls of glass face the tarmac. Planes jockey for position. Airplane noses, wings, and tails approach and recede. Outside the floor-to-ceiling windows, Ground Control directs the planes. Headphones on their ears, orange sticks glow like candy corn, they gesture to the pilots to steer the hulking steel tubes. This way, that way.

They're boarding early—I haven't even taken the pills yet. I swallow two teensy white tablets as we walk toward the boarding agent. The printer had stopped working, so the agent ripped my handwritten ticket, passing the stub to my outstretched palm like a concierge at a MoviePlex. The passenger boarding bridge slopes toward the plane and represents the tenuous separation of safety and risk. Three years since the collapse, I walk the plank and cross a threshold. To board the steel tube, I travel another bridge.

Waiting for Lorazepam to kick in, I deal with the situation silently, invisible to everyone except Rachel, whom I confide in occasionally.

I hear a friend in my memories. "What'd they give you?"

"Lorazepam—it's like generic Ativan."

"Oh, the pams. Diazepam, clonazapam. The pams are fabulous."

I see out the obstructed window view from my middle seat, the dark. The plane makes a forty-five-degree turn onto the long path of the tarmac, its edges lit like houses this time of year.

"We just turned. We're about to take off," is all I manage to squeak out to Rachel. I'm thinking too much. This is the pam's job, to stop my mind's pirouettes, invisible revolutions, laps in a hamster wheel.

A person who has seen things fall apart—and I mean really fall apart—my primal fear is easily accessible. Clinical words I heard after the collapse return. Blunt force injuries. Flying is like the poker of travel—you go all in. The logical mind knows. If a plane "collapses," everything will be lost. I ward off these thoughts. A moment before, perfect inane thoughts had occupied my consciousness. My senses are like a child to be bargained with. Wailing nonstop, a toddler in the seat ahead moans, "I don't want to." His hysterics grow louder, then wane. My senses are like that boy's—they had to be cajoled. Taxiing down the tarmac, toward the runway, the flight attendants close latches and turn off all the lights in the cabin, except for a slim cord along the recessed ceiling panel. They emerge from a small passageway with cabinets and cubbies whose contents I'm not privy to. I close my eyes, feel my chest begin burning. I take a deep breath, adjust in my seat, pull the safety belt just a bit tighter. The airplane's wheels roll over seams in the pavement and make a thunk-thunk sound. A trill and whirligig of engines, and I fight panic.

I whisper to myself: "You must trust. You must not be afraid." But I am. So I lean on what I've learned.

I think about research I did about how a plane flies (Science.howstuffworks

.com). I'd found a simple diagram of an aircraft with the forces of physics illustrated with arrows. How acceleration will launch us forward. Thrust does what again? I can't remember. Think. It was a force depicted with an arrow pointing forward from the nose. Weight was represented with an arrow pointed down. Lift with an arrow pointed up. Drag with an arrow pointed backward, rolling off the plane's tail like raindrops cascading from a groove in a leaf. These forces worked in chorus.

I tell myself, "You aren't in charge of this. It's good." (*Repeat.*) "Please. Let it be good." (*Repeat.*)

We hurl down the runway. Tires smack past pavement cracks. They ding staccato in the cabin, and the force of motion presses our chests into the upright seatbacks. I close my eyes, inhale more thoroughly than normal to calm myself. The niggling protests of that child in front sounds more possessed the longer they go on, but it's the protest of the child within all of us, the grown-up who can't wail and fuss. The seams disappear, and we experience a sort of weightlessness. My body lifts in concert with the massive 747. I'm torn about whether to slide toward the pams and nod off or stay alert. Fight it, the way Lucy fights a tranquilizer during longer trips in the car. I think, "What a pair we are, she and I." Soon after thinking this, I figure out I must close my eyes. Between the normalcy of the flight attendants at the ready, the pretty contours of the cabin, the facade of the cockpit, the orderly outlines of an intact airborne plane, it's a moment that looks too good, which means it could end at any moment. I fight visions of *Fearless*, a movie starring Jeff Bridges in which he survives an airplane crash. Everything comes apart, and people's belongings become wreckage. I fight my imagination: scenes of papers, the insides of bags, clothing, books, computers, flown apart by physics. Objects in motion stay in motion.

When people walk the aisles, I feel in my feet the floor move. I think: "They're going to break the spell. Their weight will make us fall." To cope, I imagine bumps are grooves in the road. Not thirty-nine thousand feet up. Don't think. Rachel writes product copy for Target.com. I look over her shoulder: "Just boot-iful. Katherine, Karina. They put the *K* in *kute*." And I need this copy about women's leather boots to keep my secure brain cells firing. Keep smiling. Just take a nap. "Think positive" isn't enough. Didn't work for Mom, doesn't work for me now, but "Distract" helps.

Boot-iful.

This time I don't tell Rachel what I see. This isn't my first flight after the collapse, and I tell myself I must get better. Later Rachel begins reading a book, so instead I ask: "What are you reading? Is it funny? I could use just a little help right now . . . can you tell me about it?"

"Oh yeah, sure. It's *The Case of the Good for Nothing Girlfriend*—a play on Nancy Drew, only she's Nancy Clue, and the characters are all lesbians. They're driven somewhere, and a car has broken down. Butch Midge is going to fix the car and . . . should I read that part of it to you?"

I say, "Yes please."

I look through increasingly hooded eyes at her book lit by the overhead light, and I sink into a world of playfulness. I smile with every explanation, and this frequent smiling is good for me. Distraction, like Amy's elevator trick, is key to coping. Knowledge equals skill and an ability to cope with triggers and stressors: tiny lifelines that people need more of.

When the drug finally kicks in, it puts me into a heavy stupor. I fall asleep sitting numbly like a block of ice. Walking up the aisle to deplane, I see the flight attendants and hear them say "good-bye" and "thank you." I want to throw my arms around them. A fitting accolade for a high-flying feat.

All of them and the pilots had delivered me and the scores of fellow passengers safely to A Gate and safely to The Ground. The papers I'd shuffled, still there. The carry-on I'd stowed, still there. In front of me a woman's blazer strewn across a seatback, the Chicos label visible, still there. I'd seen all of these items blown to bits. The moments before taxi, the backing up of the jet from the terminal, I'd fought visions that did me no good. Imagined scenes from the movie *Fearless*, but that didn't happen. We took off. Launched thousands of pounds into the sky—propelled across the map, sliced down the middle from south Texas to the Midwest, to Minnesota, home to my snow, through the briquette night—to a landing two hours and forty-five minutes away, where the gates would open and the 16-degree atmosphere would say, "Welcome." In my mind I'd kiss the ground, kiss the floor at home, as I visualized thanking those flight attendants. Please, just get us there. And they did.

Instead, I walk by, smile politely, and say, "Thank you."

162

My collapse friend Caroline lost her mother, Vera, and brother, Richard, in the collapse. During one of our first meetings, at TGI Fridays in Bloomington—adjacent to a busy freeway, Highway 100—Caroline and I sat outside at a high table in a fenced patio area, on a sunny day. To the low hum of constant traffic, we talked about the collapse.

Caroline said, "I wasn't sure what to expect of our meeting."

"Neither was I," I laughed.

She wasn't sure if I'd want to talk about it. I told her, "I wasn't sure what you'd want either."

"If you'd want to ask lots of questions—and in how much detail? And when you looked at me, would you think, as I was saying something (anything: what was hard, what hurt physically, what I worried about, thought about), would you think: I hate her? I wish it was my family here instead of hers?"

Would she feel the loss so completely that it would be hard to be around me?

Caroline recoiled and said, "No. No. No. I don't feel that way at all."

I felt my body relax, my shoulders lower. We talked about how she's recurrently sad, how she cries often. How she breaks down in front of her friends, how they treat her like an alien, in a way, because they don't know what to do, how to help, and they want her to be better. They want things to be normal, like they used to be, where they could go to a bar, socialize—drink, eat, laugh, gossip, talk about celebrities, or complain about mundane household things like laundry not getting done or the weather being bad. I nod over and over, the sense of recognition strong and building momentum. Though we went through such different situations that day, our grief and life feeling shattered changed us.

She asked me if I think about the bridge collapse every day. "Yeah," I said, nodding, "I think so . . . pretty much every day. Because if I didn't, I'd wake up the next morning, and I'd known I'd gone a day without it, a day of forgetting." I said, "That hasn't happened yet."

I told Caroline how I have a prototype of the bumper sticker I'd designed originally, the one with all thirteen names, on the ceiling of my car.

"Something in me died that day with them," Caroline said. I nodded again and felt so sad for her.

Caroline said simple things too.

"I just want them back."

"Of course."

We talked for four hours. The medical examiner said Caroline's mom, Vera, drowned. Tears filled Caroline's eyes. "That's something that's hard to think about," she said, to think her mom suffered some. Her younger brother, Richard, died from blunt force injuries to the head, probably in four seconds or so, but worse, his body was found in pieces—Caroline speculated, from the sharp steel and the impact of the collapse. Their car was the deepest in the river, with the most cars and steel on top of them. When found, Vera's arms were wrapped around Richard.

Caroline wanted to know, what was it like—the fall? As horrible as her losses were, knowing was better than guessing. She didn't see their bodies, was advised against it because they had been in the water too long. Not knowing, the mind can run marathons: metaphorical trips up stairs, over miles of freeway, down gullies, into puddles and hallways and stale rooms, through trains and past children on playgrounds, past mail carriers and the public. The mind can guess and exhaust itself.

Vera had been a refugee and had immigrated to the United States from Cambodia during the Vietnam War. Vera talked to Caroline all the time about her hardships there. She walked with bleeding feet, with everything they owned in their arms. They endured starvation and lived with the constant fear of stepping on mines. In the United States, in Minnesota, Vera worked twelve-hour days in a factory. Vera escaped Cambodia on August 1, amazing coincidence, 1980, and was seven months pregnant. Twenty-seven years later, when Vera was killed in the bridge collapse, it was August 1, 2007, and her daughter Caroline was seven months pregnant with Vera's grandchild, Brady.

During the time when the survivor compensation process was becoming law, Caroline was at a mall, nine months pregnant. Out in the parking lot, a woman approached and said to Caroline, "I found a wallet," and as she opened it, cash began to tumble out. The woman asked if she could get in Caroline's car. This is where Caroline stopped during the telling, with a quirky embarrassed smile.

"I'm really gullible," she said.

Then a second woman got in the car and pulled a gun on her. They knew she had lost her mom and brother and said, "We know you have money." Caroline told them, no, she didn't, and at the time that was true. Our cases hadn't even gone to the lawyer panel yet. The two women said to Caroline, "It was in the paper." The women took Caroline to an ATM, where Caroline proved she only had five hundred dollars. The women took all of it. They were never caught.

In the car I buckled up and, with my mittened right hand, touched the names on the sticker on the ceiling. Driving home, I tried not to cry. The farther I drove, the less important that seemed. Why not cry for my new bridge friend? There was no one here who I had to be strong for. No one I had to hold it together for. So I cried. I let it come in waves, naturally. I told Caroline many times that I've been learning so much since the collapse. I talked about my work with Amy, how she believes that emotions and the body are intimately intertwined and if feelings get "stuck" in the body, they can make you ill. I told her about a book I found and highly recommended—*Letting Go of the Person You Used to Be* by Lama Surya Das. I let the tears fall. When I turned onto my street, I could hear the ice crack beneath the tires and wondered how it could feel so long ago and like yesterday all at the same time.

163

On the nightly news they're talking about Sully Sullenberger, a pilot who landed a troubled airliner on the Hudson River after a double bird strike. I can't help myself. I think, How lucky they are in a way . . . the crew and passengers of flight 1549. No one died. No one has to have survivor's guilt. The feeling that aching should be something for which you are grateful.

I use the posture pump at night and continue with therapies. Sometimes I get headaches, sometimes there is pain, but there's also progress. There's much to give thanks for, much that is good now. There are times when I feel angry, feelings come sporadically. But I'm trying to forgive. It's a process—learning to let go and trust again. Worrying about the others was a way to speak up instead of being silent, like I had been for Mom. I'll always be proud of that.

Barb demystified EMDR. Going to therapy doesn't mean you're weird or weak—it's a smart strategy toward recovery. Amy demystified massage: it's not just a luxury—it's preventive medicine for well-being, like taking vitamins. And yoga isn't just exercise. It brings you into the depths of your body's workings—mind, heart, spirit. Trauma can get stuck in your body and wreak havoc. These therapies are both restorative and preventive medicine.

164

After another interview a videographer asked me to run the mic cable up through my shirt, an act I and other survivors had grown accustomed to. He attached the mic to my collar with a clip. Afterward he said he was impressed.

I asked, "Why?"

"Because," he said, "I heard your heart beating. Steady. One dependable beat after the other. You were talking about this terrifying moment when the bridge collapsed, but you were cool as a cucumber."

165

Telling this story has been a bridge. One that's helped me piece together what falling has meant in my life. Friends teach me new lessons, help me laugh, see the world renewed. Other lessons arrive without major incident.

"Kisses are great," I said, as Rachel and I leaned into the kitchen counters. We realized I process out loud. Regularly, there will be something I say instead of feel. Rachel said: "Don't say so much about how great it is. Just let it be great." We sipped ice water and stared. We accomplished something big, and we knew it. I must still work to not overthink. As I learn to reconnect with my body and let go of my logical mind, I realize it's a work in progress. Instead of kissing like we're on a deadline, we slow down and linger with no place we have to be. This gentleness, this quietness, connects us.

Leaning against the kitchen sink, I hold her, feel the bones in her chest, her steady body in my hands. I encircle her in my arms. Tears fall between my cheeks and glasses, and I say after a pause, "I don't want us to ever die." Then all I can do is squeeze, knowing my wish can never come true. Holding is all I can wish for. We mark another moment, a good one, of life.

When I drive home with Rachel—whatever the event we're coming from, doesn't really matter—there's a song by Michael Franti that speaks to us. It's track 13 from the album *All Rebel Rockers* called "Have a Little Faith." Even though the freeway is a quicker, more direct route home, at times we opt for side streets, turning left by Lake Nokomis so we can circle it.

Michael Franti sings, "I know it's hard / when you are down."

The bridge to the song plays, and we go quiet, riding together, windows down, cool evening air caressing our skin.

And these are nights when we replay track 13. We see our lives before us and feel the miles behind us and feel the gratefulness that comes, circling the lake. Faith in each other, strong for each other.

166

The collapse hints at a multitude of invisible issues neglected over years and decades. We have a vast sprawling network of human-made infrastructure that demands our attention. I feel pessimistic that anything will change, but the side of myself that wakes in the morning, ready to greet a new day, heartens me with words of encouragement. I struggle to trust. I struggle to accept that the needed maintenance will occur and protect us, innocent citizens, from the rain, wind, cold, corrosion, and effects of time, shearing away strength from that which should be strongest. I struggle to believe in concrete and rebar, but unless I get an education as a bridge inspector, I am in a similar situation that I was in pre-35W. I am an ordinary citizen, a taxpayer, unable to inspect bridges myself. I must believe in the conscientious women and men who work hard to keep us safe. I must believe that the next 35W collapse will be averted. I must look to the sky and cross my heart and pray that there will be no more next time. But no matter how I ruminate over broken bolts, it cannot change the fact that they stayed broken. How many bolts are broken all across our state? How many bolts, across our nation, remain missing from the positions where they're supposed to be, securing bridge members, reducing shear, making a bridge strong? How many remain broken? We have hundreds of thousands of bridges. How will we ever know?

167

Flash to 2011, to a reading series called "Readings by Writers" curated by local St. Paul poet laureate Carol Connolly. It was mid-November, and one of the scheduled readers that evening was Becky Lourey. From Connolly's newsletter, the *Lookout*:

"BECKY LOUREY—when serving as a respected Minnesota State Senator, was the author of the legislative Petition against War in Iraq, signed by many of her colleagues. It said '. . . We give our unyielding support to our young men and women serving in our nation's military, even if we oppose the policy that sent them to Iraq.' Becky lost one of her twelve children, Matt Lourey, a military pilot, in the Iraq war."

Becky read the full text of her letter opposing the war. She read with purpose in a slow, serious, and deliberate manner. I felt a hint of giddiness, worried she would leave early or by some circumstance I'd miss my chance to meet her. I scanned the crowd regularly, checking to see if she's still there. When the reading ended, I beelined to the other side of the room.

"Hi Becky. I'm Kimberly Brown, the bridge collapse survivor."

"Oh yes! Yes! How are you? So nice to meet you."

She wasn't what I expected. (Why do we expect anything, and what makes us think we're even close to knowing?) She's moderate height, with a short, brunette bob. Was she wearing glasses? Tiny round lenses? Vivacious and full of energy, she grasped my arms and squeezed with warmth. The strength in her hands surprised me, and when she squeezed again, I noticed her biceps, the outline of them, through her cotton sweater. I congratulated her on the power of her petition. I thanked her for her political work, for writing to me. The conversation was brief, but when I turned, I realized a clump of people had formed, waiting to talk to her. When we wrapped up, she wished me well, hoped I was healing, and asked me to stay in touch.

168

I lie on Amy's massage table. I smooth the fabric on the cradle and then place my face in the circle. *Close my eyes. Deep breath. Re-lax.* Amy worked fingers and palms over my back. Tips of hands wiggle a line up and down both sides of my spine like salmon swishing upstream, a solitary finger traces a freeway path across my city. L5 aches like a bruise. Movement, stretching, rips. Amy pushes near my shoulders and lower back in opposite directions. *C'mon L5, lengthen*, she tells it, like she's holding open a door. *Please. Do. Come in.* Lengthening. Like a girl chewing bubble gum, pulling it longer. Rubber bands stretch L5. Give it more space.

Music—long lines of bells, chords, and notes—plays. Fingers press suppleness. Back muscles amaze. Alchemy of warmth and pressure, Amy presses health back into my body. Images pop into my mind.

Imagine a different ending for Mom. She is on the table instead of me. This massage is something she wouldn't have done. Might I have convinced her had she, had we, had a second chance? My form swaps out, and Mom's swaps in. I hover like a spirit above the table. She lies face down, her body warm. Her long limbs are different, better. Her back isn't dry and flaking, itching severely from all the meds. No need this time for her to ask, "Will you scratch?" I don't watch flakes of skin fall like a cascade of instant potatoes. Instead, she's youthful. Soft, supple skin. Blood flows life into cells and veins. No pain. No more low high low high blood sugar cycles. Quiet, slow, steady pressing of health back into her. She lies, eyes closed, on this table, her body absorbing skilled hands that press through knots, calm nerves, stretch muscles—hands that will her to heal, to live.

My pulse steadies. It'll all be okay.

Flash! For the first time I imagine myself older, hitting milestones. Forty, fifty. Twenty years out and longer. The collapse lasted only thirteen seconds. There are other griefs that are behind me now. It is time to embrace this sudden image. This sudden image that I'll live, that I have time.

169

It is winter in the Midwest. She rarely refers to herself as a survivor, something about that label feels odd, yet she knows that she is. These days she is years from the trauma that nearly took her life. These days she likes to go out to the hard surfaces of her property to do a sort of meditation. She never understood why people love so much to golf or fish, but shoveling brings her closer to understanding the effect of time spent repeating an action. It's been a slow start to winter, but lately it's been snowing nearly every other day. On the other side of the railing, the long silver handle of the wide red shovel waits. The Ace Hardware label fell off long ago, and the blade is no longer straight. The edges curve slightly, the result of efforts spent hacking away at ice. The past two snows began as rain, which turned solid as temps plummeted, then the waterlogged coating of snow gripped everything in its path, including the bare naked arms of trees reaching out in all directions.

"Snow blowers are for sissies, she's a shoveling machine," the wife posts to Facebook. The sound of the scrape echoes off the brick house. The goal is to reach pavement, which she gradually will, as she moves in chop-chop motions all around, as she patiently reveals that which was hidden. She is sweaty and not more than ten minutes into it. So warm, she unzips her down-filled parka and removes her gloves. She's pushed the snow from the driveway, all the way down to the pavement in part of it, chipping at the hard scab until the middle gives way. She shovels loads that are heavy-wet, but she's not picking her body, a compulsion she fights. This is the ideal exercise for tackling anxiety, for a girl who, after surviving, isn't sure what to do.

People complain about the weather. It's a dependable topic of conversation, to be Minnesota Nice but avoid any real intimacy in thought or conversation. It's been years since she tumbled—far, far from the sky—when the road gave way and it seemed certain it was her time to go. When it wasn't, there were invisible consequences to tackle, and now she fights to be calm; she works to feel peace. Now she reaches the edges of the cleared pavement, and she lugs the loads that precipitated overnight when all the people of the neighborhood closed their eyes and slept. Now she relishes the strength in her back, even the pinch in L5 that remains—even that is welcome. The ache of her shoulders proves that she is not imagining survival. The slipping beneath the shovel as she jags at the ice lip with her left foot results in a series of satisfying soul-loving crunches. Ice flies and acquiesces. She pushes again with her left foot braced on the shovel, and she feels how her right buttock flexes, and she feels as she splits forward how limber her warmed body is now working. A machine indeed. She requires no octane, no gasoline. She runs purely on love and wanting. Lift with the legs, walk the gleaming pavement, glistening as if exalted via ritual lavation. Her body pleases her, thoroughly from her hands on the shovel's handle, up the arms, over the flexed neck muscles, down the back, and all the way to her lined boots, where her feet stay dry and protected.

The blisters on her hands—a latest allergic reaction to stress, it's all the doctors can conclude after the labs come back negative—the blisters are healing. This is how not knowing what to do manifests physically. This is how an over-awareness of time materializes in something visible. She remembers past winters after the accident when shoveling was almost impossible. Her body frustrates her and thrills her equally. She no longer questions why but clomps to the garage. Her brow wet beneath her knit hat. She scoops Morton's rock salt into a sack, clomps back, and then sprinkles it across a stubborn swath of immovable driveway ice, compacted by wind chill and automobiles led into their garage slots. She sprinkles the salt in loopy arcs, and it pops already, landing there. The melting will return the driveway to its original glory, and this pleases her. This is something she has done. She has created this cleanliness, this order, this magnificently travelable pathway. The expanse will dry. Sans ice, sans snow. Ready to welcome parked cars of all stripes: of friends, of family, of the neighbor who needs to change direction in the alley.

The sidewalks, the patio, the front steps all around, are clean, like a scab she can finally pick and discard. Healed. A whole and polished surface: all that's left behind.

170

Rachel is my minor historian. I put some older pictures from 2011 up on the bulletin board. Rachel remembered.

"That's when you started having skin stuff."

"It was?" I'm shocked. "That long ago? How do you remember that?"

"Yeah. We were sitting beneath the picnic area canopies eating with Katie and Richard's people, and you were showing them your hands."

I nodded, memory returning.

"You didn't know what it was then. Thought maybe they were warts, so you had started treating them for that, but that's not what it was."

"Yeah," I said, remembering. There were the hives about three weeks after the collapse. I had thought that was it.

I had a horrible skin problem in 2014 and the year before. It wouldn't get better. I went to the doctor, who told me to go to the dermatologist. They prescribed what they'd normally prescribe, which was essentially: take Benadryl every night, take Claritin during the day, and use this steroid cream. Oh, and by the way, yours is so bad, take it all the time.

So I kept using it, but it wasn't solving the problem. It would suppress whatever was happening, and then more sores would pop out in different spots. I was really basically disfigured.

Anyway. For months I was sort of hiding. It was a bit of an ordeal. Then I was scared something was really wrong with me. I worried, Is something wrong with my liver? Or whatever? So I got many different tests done. And then to get these tests done, I had to have a referral. It was a big mess.

So I got tested for lupus. I don't have lupus. My liver function is fine. My kidney function is fine. White blood cell count normal. They couldn't find anything. They concluded my case was idiopathic.

Idiotic is more like it.

So I felt like, *Why?* Why the hell does my face look ... just terrible? Long story short, I found a homeopath, Teresa. And I had doubts. I thought, I don't believe in that. Whatever. I mean, *maybe* I kind of believe in it. Maybe?

But I kind of don't. But I'll try it anyway. Because what else am I going to do? Nothing's working.

Through a long, convoluted process, Teresa basically told me that my skin stuff was a result of 35W.

What?! No! It can't be!

Maybe it was a stress response. Mind-body? I couldn't wrap my mind around what my body was doing or how it had an accumulation of built-up stress within itself from trauma that happened years ago. So I gathered my resources, my strength. I embarked on a twenty-one-day detox. I read articles to attempt to understand what was happening. I quit caffeine. I ate clean. She started me on something she called a "remedy," and she suggested several homeopathic creams. She kept telling me that whatever was in my body that my body didn't like had to come *out*, and the steroid cream, cortisone cream, and other drugs just pushed it back *in*. These were concepts I'd never heard of.

I didn't understand what caused this. Teresa the homeopath kept saying that when she prescribes a remedy, she's trying to find something that will support me.

Support me? How can tiny white pellets support me? I didn't understand, but where else was there to turn? So I went forward with her remedies and her advice. I went forward with faith that this time it would heal me. Even farther than before, I trusted. I told myself, *You can do this.*

So I picked up a third remedy when my facial eczema returned. I had been battling it every day. I tried to forget how miserable it made me until I had to be somewhere or see someone I knew. Misery was a word that came to mind, but was this too strong? It was a presence, this irritated skin. The burning was enough to make me want to claw my face off. And my hands—my poor hands. They were bad too. Every day, especially at night, but not just then, it swelled. It was so hot. Was there relief anywhere? In the end I clawed at spots. Vicious pathetic circle. I was scarred scarred scarred.

After finishing the remedy, a friend asked, "Did this work?"

"Yeah."

"Oh my god."

"It was a miracle."

"Is it gone for good?"

I said I thought so, most of it. I still had spots on my hands. I had scarring, some itchiness. I still had some faded spots on my face and along my jawline that were visible to me, but maybe others wouldn't notice.

171

On the promise that I'd buy lunch, I roped Rach into taking the four-hour round-trip ride with me to the guardrail and the pier. I stood on the slushy embankment and took pictures of the repaired guardrail to the soundtrack of cars and trucks lumbering past at fifty miles per hour. In 2014 I saw that the New Ulm guardrail had been repaired and the piers under I-394 were smooth and sturdy. The straps were gone and the corroding concrete spall repaired. No more rusting steel, expanding and making mini-fissures where more water and salt can enter. In 2015 several MN/DOT projects were under way, including construction and replacement of the Cayuga Bridge at I-35E.

19. Repaired guardrail; Hwy. 15 Bridge 9200, New Ulm, Minnesota. © 2014 Kimberly J. Brown.

Repair projects were up and running all over the state. That year Governor Mark Dayton announced plans to invest the state's budget surplus in kids, working families, and the state's long-neglected infrastructure needs.

Did they do this because of my letter? A friend asked me if I thought these repairs were because of my actions. I don't know the answer. I suppose it's possible. It suggests that someone may have read my letter and got busy. I have no

20. Repaired pier under I-394. © 2014 Kimberly J. Brown.

21. Another repaired pier under I-394. © 2014 Kimberly J. Brown.

22. No more straps; repaired piers under I-394. © 2014 Kimberly J. Brown.

way of knowing, but as a citizen, it is the best outcome I could hope for. Maybe I'm enough of a pain in the neck that it's easier to just fix the bridges. Whatever transpired, my feelings can be summed up in one word. *Hallelujah.*

When I started looking at bridges, through trial and error I figured out how to find the inspection reports for my local Department of Transportation (DOT). You can find these inspection reports too. The best results I had was by starting in a browser and typing into my chosen search engine "Minnesota Department of Transportation inspection reports," or navigate to your state's DOT website. Minnesota's is http://www.dot.state.mn.us. Any bridge or overpass will have an assigned number or name. You may need to begin with the intersection and narrow down from there. For example, to find the New Ulm guardrail, I looked for Bridge 9200 inspection reports. If you have questions or need help, contact your state's department of transportation.

In November 2014 I discovered that the bridge inspection reports were no longer available in a freely accessible searchable database and couldn't be accessed without a password. This wasn't right. These inspection reports should be public knowledge. Our taxes pay for these bridges. I immediately sent a request to MN/DOT asking if I'd missed the link somehow, asking, Where was the access? Anyone who wants to look up an inspection for any bridge used to be able to via a searchable database on MN/DOT's website. I contacted MN/DOT, and the agency restored public access.

172

Pittsburgh, Pennsylvania, has the greatest concentration of structurally deficient bridges in a metro area, according to ASCE, as highlighted in a *60 Minutes* report, "American Bridges in Need of Repair," by Steve Kroft. How do you keep a bridge from collapsing? Maintenance is the answer.

On a February evening before Valentine's Day in 2016, Rach and I settled onto the couch to watch a movie, *Truth*, starring Cate Blanchett as producer Mary Mapes and Robert Redford as anchor Dan Rather. Strengthened by seeing Mapes's side of the story, as the end credits rolled, an unexpected intuition overcame me. Was it really true—the cause of the bridge collapse—was it really as simple as "a design error"? In my research had I missed something?

Like a detective on a cold case, I returned to my bookcase, hoisted a thick binder from the bottom shelf. Inside its covers, in date order, were carefully organized ephemera. I haven't looked at its contents in some time. In fact, just opening the cover causes a knot in my stomach. In a pocket I'd stashed REMEMBER THE 13 bumper stickers and various pamphlets with titles like "What is EMDR?," "Subluxation," and "Hives (Uticaria)." I turn the pages. There's a stack of I-35W collapse documents produced by the lawyers, addressed to the Minnesota Data Practices Compliance Official. Bound with a metal clip, they sit loosely inside the front cover (no space to fit them in the three rings). There are personal items: letters of support from coworkers and friends, programs from survivor benefits. I keep turning pages. The binder contains a volume of documents amassed since the collapse: inspection reports; diagrams; email and phone lists; printouts of television stories; newspaper clippings; emails sent and emails received; handouts from the legislative hearings at the Minnesota State Capitol; an instruction flyer for resqme, a pocket-sized tool that can break automobile glass in case of an accident. A few pages from the end, I find an attorney-client privileged memo dated March 25, 2009. This letter discusses the consortium expert's preliminary conclusions on the cause of the bridge collapse. Here I was. Nearly six years from the date of the letter. I go to the consortium's website and see its expert's conclusions. I'm stunned and amazed. Here is proof! Oh my god.

As the reality of what this conclusion means begins to sink in, I slip into a lasting depression whose weight crushes me. This confirms what the evidence has been teaching me all along: that the bridge collapsed not only due to an anomaly or a design error but from lack of maintenance. All of the people who are hurt, some whom I know and care for; myself, Rachel; the thirteen people who died—here is the proof. It could've been prevented.

173

I watch a nature show on TV. In the Pacific Ocean warm air heated by the sun evaporates water and creates cyclones, about fifty a year, some around five hundred miles in diameter. Near weightless spiders are carried up, up, and away, into the winds of the atmosphere, over hundreds of miles of ocean.

Larger insects like butterflies, bats, and even birds or lizards are taken up into the circular vortex and are flung out to sea. Many don't survive. But the few that do, that are lucky enough to land on solid ground, live to fulfill an unexpected destiny. The narrator said, "And from these survivors, a whole island dynasty may be born."

Could our human plight, those of us injured on the bridge and all of the people who encircle us—all the people in the state and region, the prayers, medical care, the collective grieving—could all of it "build a dynasty"? It felt far-fetched. But at that juncture of life, to the tick-tock of the universe's clock, I wanted to indulge that dream, imagine it might be possible, and assume there could be renewed strength and perseverance in that hope.

174

So why did the bridge collapse?

Pay attention. I'm going to tell you.

On behalf of the consortium of twenty Minnesota law firms, representing more than ninety victims and their families on a pro bono basis, the consortium retained some of the best experts ("Experts") in the world, who disagreed with the NTSB's conclusion.

Turns out, I had this information all along. It sat in my closet—in a memo from the consortium dated March 2009. So overwhelmed by all the information coming at us, trauma and grief, I just didn't, couldn't even, digest it. But distance, time, maturity even, and an ability to find this enhanced voice, I understand better today than I ever have what caused the bridge to fail.

The first was Professor Rene Testa, who is the former chairman of the Engineering Department at Columbia University and current full-time professor at Columbia. He has extraordinary experience in failure analysis with an emphasis in bridge failures as well as rehabilitation and remediation. The other experts were Thornton Tomasetti, a worldwide structural engineering company of 1,200 engineers, architects, and sustainability and support professionals based in New York City providing engineering design, investigation, and analysis services to clients worldwide on projects of every size and level of complexity. They were one of the lead forensic investigators on the

World Trade Center disaster. Among many other notable projects, they were involved with the construction of the three largest buildings in the world.

From Thornton Tomasetti's "Annual Report 2007": "Following the tragic August 1 collapse of the I-35 west bridge in Minneapolis, which injured 145 and left 13 dead, Thornton Tomasetti began an investigation into the cause of the collapse. This includes on-site observations, review of more than 50,000 documents, and detailed analysis of the entire bridge." The company's work involved the creation of forensic computer animations that gave engineers a view of each bridge component. They looked at photographs, video, design and shop drawings, as well as forty years of maintenance reports.

In 2007, in anticipation of the one-year collapse anniversary, when the NTSB conclusion would come due, the experts embarked on an analysis of the bridge failure in an effort to determine cause. Their findings, while not widely known, portray another set of circumstances that tell me this bridge failed due to a perfect storm of inadequate maintenance, which goes beyond the NTSB conclusion that all but exonerated neglect and the underfunding of America's collective infrastructure needs.

The NTSB concluded that the bridge fell because an "underdesigned" gusset plate was too thin to support the bridge load—including the weight of traffic, the concrete deck with added thickness, and the weight of construction materials that happened to be stored at the time over the susceptible and weak gusset plate.

First, some basic bridge building concepts.

One, steel expands and contracts. In heat steel gets bigger. In cold it gets smaller. Bridges are composed of unwieldy parts and are overdesigned to carry loads of traffic and weight. To endure the stress of cycling climactic conditions, there are certain parts that are designed to allow a bridge to "move" to accommodate fluctuations in temperature that cause steel to, in a sense, change shape (changes invisible to the naked eye).

Two, there are parts of a bridge that allow a bridge to move. The parts that are relevant here are the huge labyrinth structure of trusses (remember all those triangles) composed of steel "members," or beams, and something called roller bearings. Roller bearings are situated atop piers, which are vertical and connect to the ground (or deep into a riverbed).

While the roller bearings can be thought of as a point, like the caster, or ball, at the bottom of a chair leg, the roller bearings atop bridge piers are actually composed of multiple roller bearings—in 35W's case, four—that are actually several feet long. You can picture them by imagining several baker's rolling pins. These roller bearings are designed to move, at most two or three inches each year (they may roll a number of times each year), to support the trusses above, as the steel expands and contracts. The entire truss labyrinth above connects to its piers first at the roller bearings and allows the structure to shift.

While the NTSB put the blame solely on that skinny weakling gusset plate, the picture that emerges when one imagines the frozen roller bearings is more sinister. When the stresses built up, as the sun warmed the southwest trusses and the steel expanded, with the roller bearings frozen—that is, rusted in place to the pier—it started a chain reaction of failures that caused the bridge to ultimately break into pieces. There was no question that frozen roller bearings atop the southern pier were a substantial contributing factor.

This condition was known by MN/DOT (even documented in inspection photos) and showed to the agency in 2004 by a consulting company called URS, which the state had hired to assess the bridge.

So what?

Allow me to translate what this means to me. One, this tells me that the collapse cause was more complicated than the takeaway we are left with from the NTSB. It was not only a "design" mistake. It was that, in combination with neglect that caused this bridge to fail. Years upon years upon decades of "deferred maintenance," repair needs documented and cataloged but ignored.

Two, this says to me that we are in trouble: as a country, as a nation, as a home to more than three hundred million citizens who traverse one in nine structurally deficient bridges (seventy thousand) every day, we are in deep, dangerous peril. How can we continue to underfund our infrastructure with a gas tax that has not been increased since 1996 or been adjusted to account for inflation despite the knowledge that this method of funding the most basic environment in which we all must live is failing and failing and failing?

And three, if we are left with a partial truth as opposed to a full complicated and rich truth, how can we know what we are up against? When I was growing up, my parents—especially my mom—instilled in us that if you tell the truth but it's not the whole truth, in Mom's eyes, it was the same as a lie. So what? you might wonder. And so I'll again tell you what this means to me. It means that there are not-so-obvious implications and consequences that will result from this conclusion. A "fluke" and a "design error" largely exonerate maintenance, whereas if we realize that maintenance was a key contributing factor, then we, as a people and a nation, will take steps toward a future that puts the focus squarely where, I believe, it belongs: on properly maintaining the giant workhorses that shape our well-being from point A to point B and everywhere in between. A great many people think that the reason the bridge collapsed was the gusset plates. This is what the NTSB takeaway has done. I say, loudly and directly, if only we had a picture of the fuller truth. If only we did, then we could be empowered to make choices based on *all* of the information.

175

Funny how the days pass. Daily life, routine and repetition, the insignificant ways we are forced to spend our time and, cumulatively taken, our lives. The drive to work, ordinary. The roads, the itinerary, the color of the grass and shrubbery, all the same. But one morning, once past the semaphore at the green light go, the curve expands into rows of brilliance . . . the trees blushing brilliant yellow like the glowing insides of a lemon. Magnificent! The way the leaves twinkle, shaking with life—and for what feels like a solid minute, I coast along the curve, the trees waving golden.

At work after lunch, I walk around a pond. Through changing auburn colors of the sumac, the bright light of October 1 sets the leaves alight from within. Out in the pond clumps of geese glide past, seemingly unmoving, except I know that the calm water surface belies the steady work going on below, the webbed feet paddling the geese forward. As a kid, on Saturdays we'd clean, and I would run a tan dustcloth over the hunting decoys. They weren't geese but mallards: mere estimates of the real birds in nature—missing their legs and feet. I watched the real birds in nature and thought, I know how your body is shaped: the succinct roundness of the head, the neatly tucked feathers, the lack of ears, the smiling point of the beak, the burgeoning powerful chest, the smooth back.

I wear a new bright yellow sweater, the same color as that glorious line of trees, a size small. I bought it feeling slightly giddy. I've not been this myself in a long time.

When I finish a lap, I reverse direction. Madonna's song "Vogue" harmonizes. "What are you looking at?" As the music looms and gains a beat, I hear the patter of notes. There's a part of me that wants to dance to the music, but that's not okay here. So I walk plainly, the way those geese swim, all mellow to the eye. But underneath I'm moving and gyrating, a living animal with untapped potential. Now I see the geese and their cute legs and wide feet swimming below the serene surface. I see the sunlight through the branches and feel the music through my body. Strike a pose, Madonna says. I step to the lyrics and feel alive. I'm really happy today. My depression of recent weeks has abated for now. It helps to be shedding the past, to be losing weight. I am back to my pre-35W weight and then some. (Finally losing! Finally succeeding! Losing weight that I'd been wearing like emotional insulation.) To have physical symptoms stabilizing, to see color seeping into everything (the lawns, the brush, the leaves, banks of wild trees along high-speed freeways, my sweater), to see wildlife in their place—swimming, to tilt my chin up and hear the canopy seem to applaud by tapping together hundreds of leaf-size palms. To feel the heat in my feet and sense the movement in my body, as I finish the circle and head back.

176

How could we allow the bridge collapse? A good reference is the book *Too Big to Fall* by Barry LePatner: "A comprehensive overview of the shocking state of our nation's infrastructure and what must be done to fix it." Infrastructure experts are unsure if it's that we are collecting insufficient money or just not spending what we collect on maintaining critical infrastructure; rather, allowing the diversion of these funds to support other interests like building new—also, we are using much less gas now than in 1996, so tax collection is less.

In a country where one political party is reduced to the "spenders" and the other as "against all taxes," we refuse to invest the necessary money to meet our obligations and maintain a fully functioning system of infrastructure. Why? Maybe because this is hidden, serious, unsexy, stuff. Maybe because half-truths get people elected, and few people have the time, wherewithal, or industrial knowledge to cut through the noise. Perhaps we'd rather put our energy and effort where we can see it, flaunt it, where we can say look how big and powerful we are. And yet quietly, secretly, beneath these bridges—where we can generally count on minimal foot traffic translating to few eyes upon the blight—our bridges, our bolts, our roller bearings, our rust, our rot, continue. It will continue across the system, and experts warn that further bridge collapses aren't a question of *if* but *when*. Should we have to gamble? Will you make it across the next bridge you cross?

177

Middle of the night, shuffle to the bathroom. Groggy, it's sort of my fishbowl moment. A moment where sound is lacking and there's a sense of comfortable aloneness. I'm years out, and still the thoughts come. I don't command them or compel them. *We don't know when our time is up. We can be in perfect condition one second, then the next gone.* Now I've become more skilled at knowing what to do with them. I notice them like I'm people watching near a bustling crowded sidewalk. They are in front of me one moment, then the next they move along. I have more moments now where I realize I'm just so happy to be alive. At this moment, this *minute*. So happy, that I don't want to do anything. I just want to *be*. No work. No waiting. No pain. No hurt. This is what's important to me: being.

178

Here's what I didn't know: I didn't know I would study them. I didn't know bridges would be part of my life. I remember as a kid crossing those hulking bridges, straining my gaze out the side window at the steel cage above. It reminded me of my friend's Erector sets. I remember passing beneath the steel—looking up, looking up.

I didn't know that I had a sense of safety. I didn't know I had one, so I didn't know I could lose one. Now the moments that pass—as many moments where I notice or stop what I'm doing—I'm hyperaware. Self-talk, babbling almost, too aware of the difference between pre-innocence and knowledge. Seeing a bridge fall on TV or watching one get destroyed in person is nothing compared with the sight-sound-movement-gravity-thud-crash-threat that my body absorbed.

This new me feels closest to understanding what a dog experiences when it senses inclement weather. Only for me there's always a storm on the horizon, or there might be, or there could be.

Still, years later my body can detect vibrations. When a floor hums ever so slightly, I stop. The collapse will always vibrate. I flash on images: birds exploding from their perches in trees; does retreating to the cover of forest. We are more animal than human.

And isn't that the nature of life? The positivity of life. The tragedy of death. There are no guarantees.

I didn't know so many things. I didn't know how in the future all the souls crossing the new, better, stronger 35W Bridge would be dumb to the knowledge of who we were, the people who fell. We would become invisible. Invisible like the undersides of bridges. Unknown. As unknown as all the others, all the other unknowable "victims" and "survivors" who would fall in line on the chart of history. All those who belong to this unfortunate club—as if you could call us that, as if we'd chose it, which of course we did not.

179

Collapses are much more common than I'd realized. In winter 2010 the Metrodome stadium roof in downtown Minneapolis collapsed. That year several other roofs collapsed due to thirty inches of snow we'd accumulated. Yet among all of this buckling, the trusses of our roof held. Sometimes I remind myself to look at expectations, to realize how much I take for granted. Even as trusses crumble, most of the time structures remain aloft.

We can choose to fill our consciousness with bitterness or with gratitude. Everything is a choice.

180

We erect high-rises. We proliferate fast food. We string information superhighways to connect people. We build borders, fences, to keep people out or hold people in. If only we could build bridges. If only we had a *Top Bridge* to match *Top Chef*. Reality television complete with drama and suspense. If only bridge maintenance workers could compete. Who could repair the most and worst problems best? Insidious parasites, hidden, invisible, beneath the bedrock of our common reality. If only.

Bridges define us and encompass our consciousness. They span distances and provide shelter in the most fundamental ways. So inherent in our lives, they go unnoticed and unnamed. There are bridges in my house, over my head, in the truss system in the attic rafters, in the elevated floor spanning the foundation. There is a bridge over the threshold of doorways I pass under to roam in my house, where I enter the space for washing clothes or renew after stresses have compiled.

According to an ASCE 2013 report card for America's infrastructure, we cross around 607,380 bridges in the United States. Aging behemoths we surmount daily, oblivious to the pulse of disease percolating underneath. If only we cared as well for the veins of our connectors, our pathways that carve us close and far, that traverse us past national landmarks, past broad prairie, gurgling stream, soaring mountain. I dream of building bridges, not stadiums. Of building stability, not celebrity. Where are we more blessed with resources, will, and potential than in the United States, where we are the number one richest country in the world, at $14,265 trillion GDP? So rich that we outpace the number two rival, Japan, by nearly $10 trillion. But we line our pockets with status. Rolls-Royces, precious gems, vacation homes. I dream of our next generation valuing the world we create. Static structures lift us over depressions, straddle natural resources, this world we love and must care for. If only we invested in bridges that would never fall again. If only.

The Millau Viaduct, the tallest bridge in the world, taller than the Eiffel Tower, opened in France in 2004. The Brooklyn Bridge in New York captured our imagination in 1883 and in 2010 was used by 140,000 vehicles a day. California's Golden Gate Bridge opened in 1933 and allowed us to cross the shark-infested waters of the San Francisco Bay. But bridges require ongoing maintenance. The Golden Gate Bridge requires a continuous yearly paint job in "orange vermillion," which protects the steel from corrosion caused by salt in the air. I Googled "bridge maintenance budgets states." The hits say, "[Insert state here] falling down on the job of bridge maintenance." We built the U.S. interstate system and covered more than 46,876 miles, but we have failed to keep pace with the upkeep. We've allowed that.

Except where federal exemptions are granted, states are required to inspect bridges on the National Bridge Inventory at least once every two years. But there's no law requiring states to repair bridges. We bemoan the burden on taxpayers, but what did we expect? We must go into these ventures with the clarity of a five-year-old, our youngest, our brightest, who know at their core—without premeditation or avarice—that the best parts of the world are the security of a bedtime story, a night-light, and parents who tuck them in. The covers pulled beneath their chin equal bedrock and a restful night with the stars glittering past the roof that shelters their dreams and the promise that tomorrow will hold discovery, play, learning, and potential. We can do this. We can choose. No child should go to bed without a parent because a bridge was in distress, its cries ignored.

We can imagine each bridge when it was new, like a baby, a perfect human being—one we'd planned for, anticipated, welcomed into the world with fanfare and celebration. Then we must imagine that baby as it grows and develops, as it enters puberty, teens, young adulthood, and so forth. We must anticipate the bumps and bruises and bandage wounds and be prepared to help them heal. Our work isn't over with the baby's arrival. It is then that the work is just beginning.

I will hang onto hope for the future that people can come together and create a better world. We have to, don't we?

181

Years later I still fight panic on planes. But in November 2014 I flew for the first time without taking the Lorazepam. (Good idea, bad idea?) I've been telling myself that the turbulence is just like a car going over potholes. Mostly, this helps. Yet the most subtle movements of the plane keep me working mentally. I fought rapid-fire thoughts, images (I see the aisle ahead, but the floor is suddenly vertical—*No!*), and pushed real ones away (e.g., the Malaysian plane that disappeared). On the flight back I realized abruptly that the sensations my body experiences sitting on that plane as it maneuvers through the sky are the closest approximation to how the bridge felt before it fell. *Oh-hhh.* (Or rather, "Duh"?) But it feels like I'm newly ultra-conscious of the connection (I've avoided flying for as long as possible, several years): *that's* why my body *freaks* out! The rocking, the lurching, the oscillating, the sounds of the engines "downshifting," the sensations—they're all the ones I felt all those years ago before the bridge deck dropped out. But after we landed, Rachel had said, "You were brave." I nodded. She said, "You were afraid, but you did it anyway."

182

I stepped into a rickety elevator. Another woman was there. She was African American, with a brilliant smile, nicely shaped hair, relaxed body language. I pressed the button for 13. This is yet another stress dream—it still surprises me that I have them. You see, it's April 2016, and the logical part of my being thinks I should be done with them. The elevator began lurching upward. This was something I'd repeated several times in the night. Going up. Going down. This time my co-rider began talking about a separate unrelated accident from earlier in the dream, where a piece of a building had fallen off earlier in the day and people's safety was in jeopardy. *Please, no*, I told her. I leaned face first into the corner of the elevator, wishing to blind my vision from reality. The dread was growing. *I have such terrible PTSD sometimes*, I told the woman. My mind showed me the elevator gaining speed and then launching out the top of the building. I flailed inside the airborne box, falling, terrified, the part where they cut away on TV or movies. As I began awakening, I shared the dream with Rach.

"Do you need a hug?" she asked me.

"No. I don't know. Well, I guess a hug is always good," I said, and as I went in for the hug, I tried to explain what I was feeling. I was terrified, yes. I was battling my body's post-traumatic reactions, yes. Yet despite this, I continued forward, entering the elevator, riding it to the thirteenth floor. "It's like, the dream is telling me that this is just what life is and what it's going to be," I started. "I'm going to keep on doing things despite being afraid. While doing them, I'm going to be terrified or uncomfortable—it's just how it is. But I'm always going to." Once awake, I was realizing that this was the epitome of courage, both in action and in practice. It's doing the unpleasant thing, the scary thing, the dreaded thing, despite every memory that urges me toward retreat.

183

With time come anniversaries. Each year, when the bridge collapse anniversary comes around, I do mark it, but it's much less heavy for me because I have been to the heavy place. And it's not good for me anymore, and it's not good for us, Rachel and I. A few years prior to the collapse, on December 30, 2005, Rachel and I traveled from our northern tundra to a farther northern tundra, to Toronto, Ontario, Canada, to a society advanced enough to understand that people come in all varieties and when love beckons one to bind herself to another, who was the government to stand in their way?

When we came home, still floating on the elated bliss of newlyweds, it was a shock to go back to our separate things: filing taxes as single persons, paying for our separate medical insurance plans. But in 2013, August 1, marriage equality was signed into law in our home state, Minnesota. Rachel and I now file taxes as married. I'm covered under Rachel's health insurance. Despite collapse, we've stayed together, and we can't imagine life without each other. August 1 is now not only a day to remember what was lost but a day to celebrate what was found.

184

After the bridge collapsed, I looked the same on the outside. But what wasn't visible was my crooked compacted spine and PTSD. But what I am about to say is a celebration: of resilience and metaphors. When I couldn't get my mind to relax during sleep, I had to find a bridge. This is my mouth guard. I've worn it every night, and now it has a hole in it where light shines through, like the Cayuga Bridge deck. Time for repair.

My teeth have abfraction (mechanical loss of tooth structure where pieces of enamel along the gum line have broken off, caused by forces applied to the tooth), but this marvel has done its job, preventing further damage. It is time for a new guard—an acrylic leaner one is in the works (thank you, Grand Avenue Dental), and a new "bridge" will become part of my every day.

On the eighth collapse anniversary, on Facebook, I saluted all my bridge friends for their resiliency and strength to overcome all the effects that recovery demands, including the invisible ones. "Happy anniversary to us," I wrote. "Stay hopeful, stay positive."

185

Memories in this life build in layers. Rising through blank space, beyond our sight line, bridges surround us. Become our stories. Portions of memory get buried, knocked loose by recall of words or phrases. I want to follow north. Let it come around south. Get lost, get found. Be able to look back at grief, as if it were an exit passed. I want to see ahead and know I chose to follow the curve though I couldn't see around it. I want to know the lemonade lights above marked my way. That the road rose and connected to the earth, rooted on a map, one we saw and tried to know, as the morning passes into rush hour and I live to cross another bridge.

Acknowledgments

I give thanks for many people. To all who worked to craft a Bridge Survivors bill for long-term recovery. My pro bono attorney Wil Fluegel; Nancy Remington; Chris Messerly; Phil Sieff; Lisa Weyrauch; Joel Carlson; Ryan Winkler, author of House File 2553; Speaker of the House Margaret Anderson Kelliher; Senators Lawrence J. Pogemiller—majority leader, Jim Carlson, and Patrícia Torres Ray; Conference Committee members: Senators David Hann, the late Linda Scheid, Don Betzold, Mee Moua, and Ron Latz; Representatives Phyllis Kahn, Loren Solberg, Steve Simon, and Chris Delaforest; special masters panel members: attorneys Susan Holden, Steven Kirsch, and Mike Tewksbury. Thank you, U.S. senator Amy Klobuchar and the late Minnesota congressman James Oberstar. Thank you to Thornton Tomasetti for the years of work to investigate the bridge collapse cause. Thank you to lead engineer John Abruzzo, PE, LEED AP—senior principal and Elisabeth Malsch, associate principal, Forensics (both of Thornton Tomasetti), for verifying the technical veracity of section 174.

Thank you to the United Way of the Twin Cities; Leanne Mairs, American Red Cross; Survivor Resources; Rhonda Prast and Pam Louwagie of the *Star Tribune*; all the other media professionals who told our stories with dignity; and all Minnesotans who worked, prayed, gave money, made casseroles, grieved, remembered, and hoped.

Writing this book has been fraught with ups and downs. I had all the tasks all writers have—but as a trauma survivor, I had the extra task of healing while guarding my stability and sanity. I'm grateful to many people for their help. Thanks to Marianne Kelley, Barbara Stamp, Amy Mattila, Nancy Boler,

Silke Schroeder, and Kris Koestner. Thanks to talented writers, authors, and teachers. Ember Johnson for writing mojo via The Legion-Hastings VFW in the shadow of the (then) functionally obsolete Highway 61 Bridge. Naomi Shihab Nye for telling me I fit everywhere. Jonis Agee for research. Toi Derricotte for revising via recorder. Mary Carroll Moore for islands. John Medeiros for staying vulnerable. Elizabeth J. Andrew for my story's heartbeat. Appreciation to Jerod Santek, the Loft Mentor Series, the Loft Literary Center, the Jerome Foundation, and Scott Edelstein. A debt of thanks to mentor Patricia Weaver Francisco: for her *Telling* words that paved the way, for empathy surrounding post-traumatic stress, for literary vision, and for cheering each attempt after years of manuscript revisions and rejections.

To literary heroes (to name a few): Mary Karr, Cheryl Strayed, Kay Redfield Jamison, the late Barbara Robinette Moss and Lucy Grealy . . . your words came with me. Thanks to Dani Shapiro for the structure of her memoir *Devotion*.

Sincere appreciation and boundless thanks to the University of Nebraska Press, its fantastic and professional staff, and to editor Tom Swanson for believing in this book.

Mom, gone three years before the bridge collapse, your absence projected against the din of politics, pain, and post-traumatic stress. Countless times I would wish to conjure you back. That I was in such danger would have upset you, but we would've grown closer. I would have heard the magnitude of the event reflected in your reactions and in the tenor of your voice. To us (your kids) you were our ally, a confidante, someone who listened to the pettiest details, the drivel of our lives, down to the minutiae of what we ate for dinner. We would've had so much to talk about. ILU.

Since finishing this book, Rachel and I said good-bye to our precious Lucy. She was fifteen. Milk-Bone cookies to Lucy in heaven for doggy kisses.

And best for last. Rachel M. Anderson. Wife, anchor, my favorite, my best friend. Thank you for your steadfast belief in my fledgling words. Everything is better with you.

To all: please care for our bridges. Find a way.

Shall We Gather

where twisted lengths of girders
    lie along the riverbank
they seem like scraps of sky
    that dropped, dragging
its birds with them

and these were people,
    unknown, loved,
who flew awhile
    (as everyone dreams to do)
in this world of falling

Michael Dennis Browne

Remember the 13

Julia Blackhawk
AGE 32

Richard Chit, son
of Vera Peck
AGE 20

Paul Eickstadt
AGE 51

Sherry Engebretsen
AGE 60

Peter Hausmann
AGE 47

Patrick Holmes
AGE 36

Greg "Jolly" Jolstad
AGE 45

Artemio Trinidad Mena
AGE 29

Vera Peck, mother
of Richard Chit
AGE 50

Christina Sacorafas
AGE 45

Hana Sahal
AGE 22 MONTHS

Sadiya Sahal, mother
of Hana and five
months pregnant
AGE 23

Scott Sathers
AGE 29

Bibliography

American Psychiatric Association. "Let's Talk Facts about Posttraumatic Stress Disorder." Pamphlet. APA Joint Commission on Public Affairs, 1998 and 2005.

American Society of Civil Engineers. "2013 Report Card for America's Infrastructure." https://www.infrastructurereportcard.org/making-the-grade/report-card-history/2013-report-card/.

———. "2017 Infrastructure Report Card: Bridges C+." http://www.infrastructurereportcard.org/cat-item/bridges/.

Blockley, David. *Bridges: The Science and Art of the World's Most Inspiring Structures*. New York: Oxford University Press, 2010.

Brain, Marshall, Robert Lamb, and Brian Adkins. "How Airplanes Work." *How Stuff Works Science*. http://science.howstuffworks.com/transport/flight/modern/airplanes.htm.

"Bridge Failures in Ohio, Minn. Linked." UPI, January 19, 2008. www.upi.com/Top_News/2008/01/19/Bridge_failures_in_Ohio_Minn_linked/UPI-48211200784393.

deFiebre, Conrad. "35W Bridge Survivor Works to Prevent Another Tragedy." *Minnesota 20/20*, September 14, 2008. http://www.mn2020.org/issues-that-matter/transportation/35w-bridge-survivor-works-to-prevent-another-tragedy.

Diaz, Kevin, and Mike Kaszuba. "NTSB Chief Softens Comments about Bridge Collapse." *Minneapolis/St. Paul Star Tribune*, January 29, 2008.

Doyle, Pat, Dan Browning, Mike Kaszuba, and Tony Kennedy. "Effects of Bridge Work Weren't Analyzed." *Minneapolis/St. Paul Star Tribune*, August 9, 2007. http://www.startribune.com/local/11556636.html.

ECM Publishers. "House Passes Oberstar Bridge Bill." *Isanti Times*, July 25, 2008. http://archives.ecmpublishers.com/2008/07/25/house-passes-oberstar-bridge-bill/.

Federal Highway Administration. "National Bridge Inventory Rating Scale." https://www.fhwa.dot.gov/bridge/mtguide.pdf, 37–38. http://www.dot.state.mn.us/i35wbridge/pdfs/bridgenspectiondefs.pdf.

Foti, Jim. "NTSB Bridge Findings Already Criticized." *Minneapolis/St. Paul Star Tribune*, August 19, 2009. http://www.startribune.com/ntsb-bridge-findings-already-criticized/33340314/?c=y&page=all&prepage=1&refer=y.

Freed, Joshua. "Minneapolis Bridge Collapse Survivors Plead for Help." Associated Press, October 13, 2007. http://www.chron.com/disp/story.mpl/nation/5211729.html.

Haberman, Clyde. "A Disaster Brought Awareness but Little Action on Infrastructure." *New York Times*, March 3, 2014. https://www.nytimes.com/2014/03/03/us/a-disaster-brings-awareness-but-little-action-on-infrastructure.html?_r=1.

Halpern, Sue. "Virtual Iraq." In *The Best American Science and Nature Writing 2009*, edited by Elizabeth Kolbert, 116–28. Boston: Houghton Mifflin Harcourt, 2009.

Haugen, Dan. "Officials Hail New I-35W Bridge and the Workers Who Made It Happen." *MinnPost*, September 15, 2008. https://www.minnpost.com/politics-policy/2008/09/officials-hail-new-i-35w-bridge-and-workers-who-made-it-happen.

Hayes, Brian. *Infrastructure: The Book of Everything for the Industrial Landscape*. New York: Norton, 2005.

Hoppin, James. "Thirteen Years of Inspections Offer the Deepest Look into the Bridge's Condition." *St. Paul Pioneer Press*, January 2, 2008. http://www.twincities.com/allheadlines/ci_7858103?source=rv&nclick_check=1.

Iowa Department of Transportation. http://www.iowadot.gov/subcommittee/bridgetermspz.aspx (site discontinued).

James, Frank. "Last Minneapolis Bridge Collapse Lawsuit Settled for $52.4 Million." National Public Radio, August 23, 2010. http://www.npr.org/sections/thetwo-way/2010/08/23/129387414/minneapolis-bridge-collapse-lawsuits-settled-for-52-4-million.

Kennedy, T., and P. McEnroe. "Phone Call Put Brakes on Bridge Repair." *Minneapolis/St. Paul Star Tribune*, August 18, 2007. http://www.startribune.com.

Kroft, Steve. "American Bridges in Need of Repair." *60 Minutes*, S47:E10, CBS, November 23, 2014.

Kroll, John. "Bridge Design Links Lake County and Minneapolis Failures." *Cleveland Plain Dealer*, August 10, 2007. http://blog.cleveland.com/pdextra//print.html?entry=/2007/08/bridge.

LePatner, Barry. "Construction Expert Denounces the NTSB's Report on the I-35W Bridge Collapse." *American Surveyor*, November 25, 2008. http://www.amerisurv.com/content/view/5582/2/.

———. *Too Big to Fall: America's Failing Infrastructure and the Way Forward*. Irvine CA: Foster Publishing, 2010.

———. "The Year of Driving Dangerously: A Construction Expert Blasts America's Lack of Infrastructure Action One Year after the I35W Bridge Collapse." *Axis of Logic*, July 31, 2008. http://axisoflogic.com/artman/publish/Article_27847.shtml.

LePatner, Barry B., with Timothy Jacobson and Robert E. Wright. *Broken Buildings, Busted Budgets: How to Fix America's Trillion-Dollar Construction Industry*. Chicago: University of Chicago Press, 2007.

Louwagie, Pam. "I-35W Bridge Survivor Vents Outrage." *Minneapolis/St. Paul Star Tribune*, October 11, 2007. http://www.startribune.com/local/11558056.html (site discontinued), http://archive.is/0Tqr1.

———. "A Survivor's Letter." *Minneapolis/St. Paul Star Tribune*, October 11, 2007. http://www.startribune.com/local/11552931.html (site discontinued), http://archive.is/0Tqr1.

Louwagie, Pamela, and J. Walsh. "The Lafayette Bridge: A Reason for Worry in St. Paul?" *Minneapolis/St. Paul Star Tribune*, August 8, 2007. http://www.startribune.com/local/11552131.html.

Minnesota Department of Transportation. "I-35W Corridor Projects: Cayuga Project." http://www.dot.state.mn.us/metro/projects/35estpaul/cayuga.html.

———. "Interstate 35W Bridge: Original Plans & Details." Fracture Critical Bridge Inspections for Bridge 9340. Reports, 1994–2006. www.dot.state.mn.us/i35wbridge/history.html.

———. "Interstate 35W Mississippi River Bridge, Minneapolis Fact Sheet—Aug. 13, 2007." August 13, 2007.

———. "Lt. Governor Molnau Asks Taxpayers and State Employees to Submit Money Saving Ideas to MN/DOT." News release, January 15, 2003. http://www.dot.state.mn.us/newsrels/03/01/14costsavingideas.html (site discontinued).

"National Bridge Inventory." Wikipedia. https://en.wikipedia.org/wiki/National_Bridge_Inventory.

Neuro Innovations. "EMDR Therapy—An Introduction." Last updated August 22, 2017. http://www.neuroinnovations.com/emdr_therapy.html.

Oberstar, Jim. "August 6, 2008, Newsletter." *Jim Oberstar for U.S. Congress: Working for You, Working for Minnesota*, August 6, 2008. http://www.oberstar.org/news/newsletter/vol3issue8.htm.

Ohio Department of Transportation. "Bridge Terms." http://www.dot.state.oh.us/Divisions/Communications/BridgingtheGap/Pages/BridgeTermDefinitions.aspx (site modified), www.dot.state.oh.us/Divisions/Engineering/Structures/bridge%20operations%20and%20maintenance/PreventiveMaintenanceManual/BPMM/glossary.htm.

Pohland, Darcy. "Saved by Inches: Pipe Falls on Car from Overpass." WCCO-TV / CBS Broadcasting, July 29, 2008 (accessed August 14, 2008). http://wcco.com/local/falling.drain.pipe.2.783032.html.

Rosenker, Mark V. "Opening Remarks at the Minnesota I-35W Bridge Collapse Press Conference." January 15, 2008. http://2008.myvote.org/www.ntsb.gov/speeches/rosenker/mvr080115.html.

"Saint Anthony Falls." Wikipedia. http://en.wikipedia.org/wiki/Saint_Anthony_Falls.

Scheck, Tom. "Cost of Bridge Collapse Could Reach $400 Million." Minnesota Public Radio, October 1, 2007. http://www.mprnews.org/story/2007/10/01/bridgemoney.

Schipper, Henry, producer, writer. *The Crumbling of America*. History Channel, June 22, 2009.

———. *Modern Marvels: Corrosion and Decomposition*. History Channel, October 13, 2008.

Sofge, Eric, and the editors of *Popular Mechanics*. "The 10 Pieces of U.S. Infrastructure We Must Fix Now." *Popular Mechanics*, April 6, 2008. http://www.popularmechanics.com/technology/engineering/rebuilding-america/4257814.

Sommer, H., and Sterling Burnett. "Repairing Bridges without Raising Gas Taxes." National Center for Policy Analysis, October 18, 2007. http://www.ncpa.org/pub/ba597.

Stachura, Sea, and Steve Karnowski. "Attorney: NTSB Is Wrong about Cause of Bridge Collapse." Associated Press, March 25, 2009. http://www.mprnews.org/story/2009/03/25/bridgecollapse_update.

"States Warned to Inspect Bridges." CNN.com, August 2, 2007. http://www.cnn.com/2007/US/08/02/bridge.structure/index.html.

"Steam Explosion Jolts Manhattan, Killing 1." NBC News, July 18, 2007. www.msnbc.msn.com/id/19837147.

"Strength of Materials." Wikipedia. http://en.wikipedia.org/wiki/Strength_of_materials.

Stump, Jake. "Survivor Recalls Plunge from Silver Bridge into the River." *Charleston Daily Mail*, December 14, 2007. http://www.dailymail.com/News

/200712140728 (site discontinued), http://www.theufochronicles.com/2007/12/survivor-recalls-plunge-from-silver.html.
"13 Seconds in August." Multimedia project. *Minneapolis/St. Paul Star Tribune*, 2007. http://www.startribune.com/local/12166286.html (site discontinued), http://archive.is/0Tqr1.
Thornton Tomasetti. "I-35W Bridge Collapse." Report by John Abruzzo and Elisabeth Malsch and Robins, Kaplan, Miller, and Ciresi, LLP. 2008. http://www.thorntontomasetti.com/projects/i35_bridge_collapse/.
Transportation for America. "The Fix We're in For: 2015 Bridge Conditions." http://t4america.org/maps-tools/bridges/.
Tull, Matthew. "Reducing the Stigma of Mental Health Care in Veterans." About.com, June 24, 2009. http://ptsd.about.com/od/ptsdandthemilitary/a/stigmavets.htm (site modified), updated February 20, 2017, https://www.verywell.com/reducing-the-stigma-of-mental-health-care-in-veterans-2797454.
Van Hampton, Tudor. "NTSB's Gusset Alam Perplexes Engineers." *Engineering News-Record*, August 10, 2007. http://enr.construction.com/print.asp?REF=http://enr.construction.com/news/transportation/archives/070810a.asp.
Wald, Matthew L. "Deficient, but Not Necessarily Dangerous." *New York Times*, August 5, 2007. http://www.nytimes.com/2007/08/05/weekinreview/05basicB.html.
Washington State Department of Transportation. "Glossary of Bridge Terms." www.wsdot.wa.gov/Projects/SR24/I82toKeysRd/BridgeGlossary.htm (site modified), https://www.wsdot.wa.gov/TNBhistory/glossary.htm.